The Design of Everyday Things

Revised and Expanded Edition

Donald A. Norman

The MIT Press
Cambridge, Massachusetts
London, England

MIT Press books may be purchased at special quantity discounts for business or sales promotional use. For information, please email special_sales@mitpress.mit.edu.

ISBN: 978-0-262-52567-1 (pbk.)

10 9 8 7 6

Printed in Great Britain by Clays Ltd, St Ives plc

The Design of Everyday Things

ALSO BY
DON NORMAN

TEXTBOOKS

Memory and Attention: An Introduction to Human Information Processing.
First edition, 1969; second edition 1976

Human Information Processing.
(with Peter Lindsay: first edition, 1972; second edition 1977)

SCIENTIFIC MONOGRAPHS

Models of Human Memory
(edited, 1970)

Explorations in Cognition
(with David E. Rumelhart and the LNR Research Group, 1975)

Perspectives on Cognitive Science
(edited, 1981)

User Centered System Design: New Perspectives on Human-Computer Interaction
(edited with Steve Draper, 1986)

TRADE BOOKS

Learning and Memory, 1982

The Psychology of Everyday Things, 1988

The Design of Everyday Things
1990 and 2002 (paperbacks of *The Psychology of Everyday Things* with new prefaces)

The Design of Everyday Things
Revised and Expanded Edition, 2013

Turn Signals Are the Facial Expressions of Automobiles, 1992

Things That Make Us Smart, 1993

The Invisible Computer: Why Good Products Can Fail, the Personal Computer Is So Complex, and Information Appliances Are the Answer, 1998

Emotional Design: Why We Love (or Hate) Everyday Things, 2004

The Design of Future Things, 2007

A Comprehensive Strategy for Better Reading: Cognition and Emotion, 2010
(with Masanori Okimoto; my essays, with commentary in Japanese, used for teaching English as a second language to Japanese speakers)

Living with Complexity, 2011

CD-ROM

First person: Donald A. Norman. Defending Human Attributes in the Age of the Machine, 1994

For Julie

CONTENTS

Preface to the Revised Edition xi

1 The Psychopathology of Everyday Things 1
The Complexity of Modern Devices, 4
Human-Centered Design, 8
Fundamental Principles of Interaction, 10
The System Image, 31
The Paradox of Technology, 32
The Design Challenge, 34

2 The Psychology of Everyday Actions 37
How People Do Things: The Gulfs of Execution and Evaluation, 38
The Seven Stages of Action, 40
Human Thought: Mostly Subconscious, 44
Human Cognition and Emotion, 49
The Seven Stages of Action and the Three Levels of Processing, 55
People as Storytellers, 56
Blaming the Wrong Things, 59
Falsely Blaming Yourself, 65
The Seven Stages of Action: Seven Fundamental Design Principles, 71

3 Knowledge in the Head and in the World 74
Precise Behavior from Imprecise Knowledge, 75
Memory Is Knowledge in the Head, 86
The Structure of Memory, 91
Approximate Models: Memory in the Real World, 100
Knowledge in the Head, 105
The Tradeoff Between Knowledge in the World and in the Head, 109
Memory in Multiple Heads, Multiple Devices, 111
Natural Mapping, 113
Culture and Design: Natural Mappings Can Vary with Culture, 118

4 Knowing What to Do: Constraints, Discoverability, and Feedback 123
Four Kinds of Constraints: Physical, Cultural, Semantic, and Logical, 125
Applying Affordances, Signifiers, and Constraints to Everyday Objects, 132
Constraints That Force the Desired Behavior, 141
Conventions, Constraints, and Affordances, 145
The Faucet: A Case History of Design, 150
Using Sound as Signifiers, 155

5 Human Error? No, Bad Design 162
Understanding Why There Is Error, 163
Deliberate Violations, 169
Two Types of Errors: Slips and Mistakes, 170
The Classification of Slips, 173
The Classification of Mistakes, 179
Social and Institutional Pressures, 186
Reporting Error, 191
Detecting Error, 194
Designing for Error, 198
When Good Design Isn't Enough, 210
Resilience Engineering, 211
The Paradox of Automation, 213
Design Principles for Dealing with Error, 215

6 Design Thinking 217

Solving the Correct Problem, 218
The Double-Diamond Model of Design, 220
The Human-Centered Design Process, 221
What I Just Told You? It Doesn't Really Work That Way, 236
The Design Challenge, 239
Complexity Is Good; It Is Confusion That Is Bad, 247
Standardization and Technology, 248
Deliberately Making Things Difficult, 255
Design: Developing Technology for People, 257

7 Design in the World of Business 258

Competitive Forces, 259
New Technologies Force Change, 264
How Long Does It Take to Introduce a New Product?, 268
Two Forms of Innovation: Incremental and Radical, 279
The Design of Everyday Things: 1988–2038, 282
The Future of Books, 288
The Moral Obligations of Design, 291
Design Thinking and Thinking About Design, 293

Acknowledgments 299

General Readings and Notes 305

References 321

Index 331

PREFACE TO THE REVISED EDITION

In the first edition of this book, then called POET, *The Psychology of Everyday Things*, I started with these lines: "This is the book I always wanted to write, except I didn't know it." Today I do know it, so I simply say, "This is the book I always wanted to write."

This is a starter kit for good design. It is intended to be enjoyable and informative for everyone: everyday people, technical people, designers, and nondesigners. One goal is to turn readers into great observers of the absurd, of the poor design that gives rise to so many of the problems of modern life, especially of modern technology. It will also turn them into observers of the good, of the ways in which thoughtful designers have worked to make our lives easier and smoother. Good design is actually a lot harder to notice than poor design, in part because good designs fit our needs so well that the design is invisible, serving us without drawing attention to itself. Bad design, on the other hand, screams out its inadequacies, making itself very noticeable.

Along the way I lay out the fundamental principles required to eliminate problems, to turn our everyday stuff into enjoyable products that provide pleasure and satisfaction. The combination of good observation skills and good design principles is a powerful tool, one that everyone can use, even people who are not professional designers. Why? Because we are all designers in the sense that all of us deliberately design our lives, our rooms, and the way we do things. We can also design workarounds, ways of overcoming the flaws of existing devices. So, one purpose of this book is to give back your control over the products in your life: to know how to select usable and understandable ones, to know how to fix those that aren't so usable or understandable.

The first edition of the book has lived a long and healthy life. Its name was quickly changed to *Design of Everyday Things* (DOET) to make the title less cute and more descriptive. DOET has been read by the general public and by designers. It has been assigned in courses and handed out as required readings in many companies. Now, more than twenty years after its release, the book is still popular. I am delighted by the response and by the number of people who correspond with me about it, who send me further examples of thoughtless, inane design, plus occasional examples of superb design. Many readers have told me that it has changed their lives, making them more sensitive to the problems of life and to the needs of people. Some changed their careers and became designers because of the book. The response has been amazing.

Why a Revised Edition?

In the twenty-five years that have passed since the first edition of the book, technology has undergone massive change. Neither cell phones nor the Internet were in widespread usage when I wrote the book. Home networks were unheard of. Moore's law proclaims that the power of computer processors doubles roughly every two years. This means that today's computers are five thousand times more powerful than the ones available when the book was first written.

Although the fundamental design principles of *The Design of Everyday Things* are still as true and as important as when the first edition was written, the examples were badly out of date. "What is a slide projector?" students ask. Even if nothing else was to be changed, the examples had to be updated.

The principles of effective design also had to be brought up to date. Human-centered design (HCD) has emerged since the first edition, partially inspired by that book. This current edition has an entire chapter devoted to the HCD process of product development. The first edition of the book focused upon making products understandable and usable. The total experience of a product covers much more than its usability: aesthetics, pleasure, and fun play critically important roles. There was no discussion of pleasure, enjoyment, or emotion. Emotion is so important that I wrote an entire book, *Emotional Design*, about the role it plays in design. These issues are also now included in this edition.

My experiences in industry have taught me about the complexities of the real world, how cost and schedules are critical, the need to pay attention to competition, and the importance of multidisciplinary teams. I learned that the successful product has to appeal to customers, and the criteria they use to determine what to purchase may have surprisingly little overlap with the aspects that are important during usage. The best products do not always succeed. Brilliant new technologies might take decades to become accepted. To understand products, it is not enough to understand design or technology: it is critical to understand business.

What Has Changed?

For readers familiar with the earlier edition of this book, here is a brief review of the changes.

What has changed? Not much. Everything.

When I started, I assumed that the basic principles were still true, so all I needed to do was update the examples. But in the end, I rewrote everything. Why? Because although all the principles still applied, in the twenty-five years since the first edition, much has been learned. I also now know which parts were difficult and therefore need better explanations. In the interim, I also wrote many articles and six books on related topics, some of which I thought important to include in the revision. For example, the original book says nothing of what has come to be called *user experience* (a term that I was among the first to use, when in the early 1990s, the group I headed at Apple called itself "the User Experience Architect's Office"). This needed to be here.

Finally, my exposure to industry taught me much about the way products actually get deployed, so I added considerable information about the impact of budgets, schedules, and competitive pressures. When I wrote the original book, I was an academic researcher. Today, I have been an industry executive (Apple, HP, and some startups), a consultant to numerous companies, and a board member of companies. I had to include my learnings from these experiences.

Finally, one important component of the original edition was its brevity. The book could be read quickly as a basic, general introduction. I kept that feature unchanged. I tried to delete as much as I added to keep the total size about the same (I failed). The book is meant to be an introduction: advanced discussions of the topics, as well as a large number of important but more advanced topics, have been left out to maintain the compactness. The previous edition lasted from 1988 to 2013. If the new edition is to last as long, 2013 to 2038, I had to be careful to choose examples that would not be dated twenty-five years from now. As a result, I have tried not to give specific company examples. After all, who remembers the companies of twenty-five years ago? Who can predict what new companies will arise, what existing companies will disappear, and what new technologies will arise in the next twenty-five years? The one thing I can predict with certainty is that the principles of human psychology will remain the same, which means that the design principles here, based on psychology, on the nature of human cognition, emotion, action, and interaction with the world, will remain unchanged.

Here is a brief summary of the changes, chapter by chapter.

Chapter 1: The Psychopathology of Everyday Things

Signifiers are the most important addition to the chapter, a concept first introduced in my book *Living with Complexity*. The first edition had a focus upon affordances, but although affordances make sense for interaction with physical objects, they are confusing when dealing with virtual ones. As a result, affordances have created much confusion in the world of design. Affordances define what actions are possible. Signifiers specify how people discover those possibilities: signifiers are signs, perceptible signals of what can be done. Signifiers are of far more importance to designers than are affordances. Hence, the extended treatment.

I added a very brief section on HCD, a term that didn't yet exist when the first edition was published, although looking back, we see that the entire book was about HCD.

Other than that, the chapter is the same, and although all the photographs and drawings are new, the examples are pretty much the same.

Chapter 2: The Psychology of Everyday Actions

The chapter has one major addition to the coverage in the first edition: the addition of emotion. The seven-stage model of action has proven to be influential, as has the three-level model of processing (introduced in my book *Emotional Design*). In this chapter I show the interplay between these two, show that different emotions arise at the different stages, and show which stages are primarily located at each of the three levels of processing (visceral, for the elementary levels of motor action performance and perception; behavioral, for the levels of action specification and initial interpretation of the outcome; and reflective, for the development of goals, plans, and the final stage of evaluation of the outcome).

Chapter 3: Knowledge in the Head and in the World

Aside from improved and updated examples, the most important addition to this chapter is a section on culture, which is of special importance to my discussion of "natural mappings." What seems natural in one culture may not be in another. The section examines the way different cultures view time—the discussion might surprise you.

Chapter. 4: Knowing What to Do: Constraints, Discoverability, and Feedback

Few substantive changes. Better examples. The elaboration of forcing functions into two kinds: lock-in and lockout. And a section on destination control elevators, illustrating how change can be extremely disconcerting, even to professionals, even if the change is for the better.

Chapter 5: Human Error? No, Bad Design

The basics are unchanged, but the chapter itself has been heavily revised. I update the classification of errors to fit advances since the publication of the first edition. In particular, I now divide slips into two main categories—action-based and memory lapses; and mistakes into three categories—rule-based, knowledge-based, and memory lapses. (These distinctions are now common, but I introduce a slightly different way to treat memory lapses.)

Although the multiple classifications of slips provided in the first edition are still valid, many have little or no implications for design, so they have been eliminated from the revision. I provide more design-relevant examples. I show the relationship of the classification of errors, slips, and mistakes to the seven-stage model of action, something new in this revision.

The chapter concludes with a quick discussion of the difficulties posed by automation (from my book *The Design of Future Things*) and what I consider the best new approach to deal with design so as to either eliminate or minimize human error: resilience engineering.

Chapter 6: Design Thinking

This chapter is completely new. I discuss two views of human-centered design: the British Design Council's double-diamond model and the traditional HCD iteration of observation, ideation, prototyping, and testing. The first diamond is the divergence, followed by convergence, of possibilities to determine the appropriate problem. The second diamond is a divergence-convergence to determine an appropriate solution. I introduce activity-centered design as a more appropriate variant of human-centered design in many circumstances. These sections cover the theory.

The chapter then takes a radical shift in position, starting with a section entitled "What I Just Told You? It Doesn't Really Work That Way." Here is where I introduce Norman's Law: The day the product team is announced, it is behind schedule and over its budget.

I discuss challenges of design within a company, where schedules, budgets, and the competing requirements of the different divisions all provide severe constraints upon what can be accomplished. Readers from industry have told me that they welcome these sections, which capture the real pressures upon them.

The chapter concludes with a discussion of the role of standards (modified from a similar discussion in the earlier edition), plus some more general design guidelines.

Chapter 7: Design in the World of Business

This chapter is also completely new, continuing the theme started in Chapter 6 of design in the real world. Here I discuss "featuritis," the changes being forced upon us through the invention of new technologies, and the distinction between incremental and radical innovation. Everyone wants radical innovation, but the truth is, most radical innovations fail, and even when they do succeed, it can take multiple decades before they are accepted. Radical innovation, therefore, is relatively rare: incremental innovation is common.

The techniques of human-centered design are appropriate to incremental innovation: they cannot lead to radical innovations.

The chapter concludes with discussions of the trends to come, the future of books, the moral obligations of design, and the rise of small, do-it-yourself makers that are starting to revolutionize the way ideas are conceived and introduced into the marketplace: "the rise of the small," I call it.

Summary

With the passage of time, the psychology of people stays the same, but the tools and objects in the world change. Cultures change.

Technologies change. The principles of design still hold, but the way they get applied needs to be modified to account for new activities, new technologies, new methods of communication and interaction. *The Psychology of Everyday Things* was appropriate for the twentieth century: *The Design of Everyday Things* is for the twenty-first.

Don Norman
Silicon Valley, California
www.jnd.org

CHAPTER ONE

THE PSYCHOPATHOLOGY OF EVERYDAY THINGS

If I were placed in the cockpit of a modern jet airliner, my inability to perform well would neither surprise nor bother me. But why should I have trouble with doors and light switches, water faucets and stoves? "Doors?" I can hear the reader saying. "You have trouble opening doors?" Yes. I push doors that are meant to be pulled, pull doors that should be pushed, and walk into doors that neither pull nor push, but slide. Moreover, I see others having the same troubles—unnecessary troubles. My problems with doors have become so well known that confusing doors are often called "Norman doors." Imagine becoming famous for doors that don't work right. I'm pretty sure that's not what my parents planned for me. (Put "Norman doors" into your favorite search engine—be sure to include the quote marks: it makes for fascinating reading.)

How can such a simple thing as a door be so confusing? A door would seem to be about as simple a device as possible. There is not much you can do to a door: you can open it or shut it. Suppose you are in an office building, walking down a corridor. You come to a door. How does it open? Should you push or pull, on the left or the right? Maybe the door slides. If so, in which direction? I have seen doors that slide to the left, to the right, and even up into the ceiling.

FIGURE 1.1. Coffeepot for Masochists. The French artist Jacques Carelman in his series of books *Catalogue d'objets introuvables* (Catalog of unfindable objects) provides delightful examples of everyday things that are deliberately unworkable, outrageous, or otherwise ill-formed. One of my favorite items is what he calls "coffeepot for masochists." The photograph shows a copy given to me by collegues at the University of California, San Diego. It is one of my treasured art objects. (Photograph by Aymin Shamma for the author.)

The design of the door should indicate how to work it without any need for signs, certainly without any need for trial and error.

A friend told me of the time he got trapped in the doorway of a post office in a European city. The entrance was an imposing row of six glass swinging doors, followed immediately by a second, identical row. That's a standard design: it helps reduce the airflow and thus maintain the indoor temperature of the building. There was no visible hardware: obviously the doors could swing in either direction: all a person had to do was push the side of the door and enter.

My friend pushed on one of the outer doors. It swung inward, and he entered the building. Then, before he could get to the next row of doors, he was distracted and turned around for an instant. He didn't realize it at the time, but he had moved slightly to the right. So when he came to the next door and pushed it, nothing happened. "Hmm," he thought, "must be locked." So he pushed the side of the adjacent door. Nothing. Puzzled, my friend decided to go outside again. He turned around and pushed against the side of a door. Nothing. He pushed the adjacent door. Nothing. The door he had just entered no longer worked. He turned around once more and tried the inside doors again. Nothing. Concern, then mild panic. He was trapped! Just then, a group of people on the other side of the entranceway (to my friend's right) passed easily through both sets of doors. My friend hurried over to follow their path.

How could such a thing happen? A swinging door has two sides. One contains the supporting pillar and the hinge, the other is unsupported. To open the door, you must push or pull on the unsupported edge. If you push on the hinge side, nothing happens. In my friend's case, he was in a building where the designer aimed for beauty, not utility. No distracting lines, no visible pillars, no visible hinges. So how can the ordinary user know which side to push on? While distracted, my friend had moved toward the (invisible) supporting pillar, so he was pushing the doors on the hinged side. No wonder nothing happened. Attractive doors. Stylish. Probably won a design prize.

Two of the most important characteristics of good design are *discoverability* and *understanding*. Discoverability: Is it possible to even figure out what actions are possible and where and how to perform them? Understanding: What does it all mean? How is the product supposed to be used? What do all the different controls and settings mean?

The doors in the story illustrate what happens when discoverability fails. Whether the device is a door or a stove, a mobile phone or a nuclear power plant, the relevant components must be visible, and they must communicate the correct message: What actions are possible? Where and how should they be done? With doors that push, the designer must provide signals that naturally indicate where to push. These need not destroy the aesthetics. Put a vertical plate on the side to be pushed. Or make the supporting pillars visible. The vertical plate and supporting pillars are natural signals, naturally interpreted, making it easy to know just what to do: no labels needed.

With complex devices, discoverability and understanding require the aid of manuals or personal instruction. We accept this if the device is indeed complex, but it should be unnecessary for simple things. Many products defy understanding simply because they have too many functions and controls. I don't think that simple home appliances—stoves, washing machines, audio and television sets—should look like Hollywood's idea of a spaceship control room. They already do, much to our consternation. Faced with a bewildering array of controls and displays, we simply memorize one or two fixed settings to approximate what is desired.

In England I visited a home with a fancy new Italian washer-dryer combination, with super-duper multisymbol controls, all to do everything anyone could imagine doing with the washing and drying of clothes. The husband (an engineering psychologist) said he refused to go near it. The wife (a physician) said she had simply memorized one setting and tried to ignore the rest. I asked to see the manual: it was just as confusing as the device. The whole purpose of the design is lost.

The Complexity of Modern Devices

All artificial things are designed. Whether it is the layout of furniture in a room, the paths through a garden or forest, or the intricacies of an electronic device, some person or group of people had to decide upon the layout, operation, and mechanisms. Not all designed things involve physical structures. Services, lectures, rules and procedures, and the organizational structures of businesses and governments do not have physical mechanisms, but their rules of operation have to be designed, sometimes informally, sometimes precisely recorded and specified.

But even though people have designed things since prehistoric times, the field of design is relatively new, divided into many areas of specialty. Because everything is designed, the number of areas is enormous, ranging from clothes and furniture to complex control rooms and bridges. This book covers everyday things, focusing on the interplay between technology and people to ensure that the products actually fulfill human needs while being understandable and usable. In the best of cases, the products should also be delightful and enjoyable, which means that not only must the requirements of engineering, manufacturing, and ergonomics be satisfied, but attention must be paid to the entire experience, which means the aesthetics of form and the quality of interaction. The major areas of design relevant to this book are industrial design, interaction design, and experience design. None of the fields is well defined, but the focus of the efforts does vary, with industrial designers emphasizing form and material, interactive designers emphasizing understandability and usability, and experience designers emphasizing the emotional impact. Thus:

Industrial design: The professional service of creating and developing concepts and specifications that optimize the function, value, and appearance of products and systems for the mutual benefit of both user and manufacturer (from the *Industrial Design Society of America's* website).

Interaction design: The focus is upon how people interact with technology. The goal is to enhance people's understanding of what can be done, what is happening, and what has just occurred. Interaction design draws upon principles of psychology, design, art, and emotion to ensure a positive, enjoyable experience.

Experience design: The practice of designing products, processes, services, events, and environments with a focus placed on the quality and enjoyment of the total experience.

Design is concerned with how things work, how they are controlled, and the nature of the interaction between people and technology. When done well, the results are brilliant, pleasurable products. When done badly, the products are unusable, leading to great frustration and irritation. Or they might be usable, but force us to behave the way the product wishes rather than as we wish.

Machines, after all, are conceived, designed, and constructed by people. By human standards, machines are pretty limited. They do not maintain the same kind of rich history of experiences that people have in common with one another, experiences that enable us to interact with others because of this shared understanding. Instead, machines usually follow rather simple, rigid rules of behavior. If we get the rules wrong even slightly, the machine does what it is told, no matter how insensible and illogical. People are imaginative and creative, filled with common sense; that is, a lot of valuable knowledge built up over years of experience. But instead of capitalizing on these strengths, machines require us to be precise and accurate, things we are not very good at. Machines have no leeway or common sense. Moreover, many of the rules followed by a machine are known only by the machine and its designers.

When people fail to follow these bizarre, secret rules, and the machine does the wrong thing, its operators are blamed for not understanding the machine, for not following its rigid specifications. With everyday objects, the result is frustration. With complex devices and commercial and industrial processes, the resulting difficulties can lead to accidents, injuries, and even deaths. It is time to reverse the situation: to cast the blame upon the machines and their design. It is the machine and its design that are at fault. It is the duty of machines and those who design them to understand people. It is not our duty to understand the arbitrary, meaningless dictates of machines.

The reasons for the deficiencies in human-machine interaction are numerous. Some come from the limitations of today's technology. Some come from self-imposed restrictions by the designers, often to hold down cost. But most of the problems come from a complete lack of understanding of the design principles necessary for effective human-machine interaction. Why this deficiency? Because much of the design is done by engineers who are experts in technology but limited in their understanding of people. "We are people ourselves," they think, "so we understand people." But in fact, we humans are amazingly complex. Those who have not studied human behavior often think it is pretty simple. Engineers, moreover, make the mistake of thinking that logical explanation is sufficient: "If only people would read the instructions," they say, "everything would be all right."

Engineers are trained to think logically. As a result, they come to believe that all people must think this way, and they design their machines accordingly. When people have trouble, the engineers are upset, but often for the wrong reason. "What are these people doing?" they will wonder. "Why are they doing that?" The problem with the designs of most engineers is that they are too logical. We have to accept human behavior the way it is, not the way we would wish it to be.

I used to be an engineer, focused upon technical requirements, quite ignorant of people. Even after I switched into psychology and cognitive science, I still maintained my engineering emphasis upon logic and mechanism. It took a long time for me to realize that my understanding of human behavior was relevant to my interest in the design of technology. As I watched people struggle with technology, it became clear that the difficulties were caused by the technology, not the people.

I was called upon to help analyze the American nuclear power plant accident at Three Mile Island (the island name comes from the fact that it is located on a river, three miles south of Middletown in the state of Pennsylvania). In this incident, a rather simple mechanical failure was misdiagnosed. This led to several days of difficulties and confusion, total destruction of the reactor, and a very close call to a severe radiation release, all of which brought the American nuclear power industry to a complete halt. The operators were blamed for these failures: "human error" was the immediate analysis. But the committee I was on discovered that the plant's control rooms were so poorly designed that error was inevitable: design was at fault, not the operators. The moral was simple: we were designing things for people, so we needed to understand both technology and people. But that's a difficult step for many engineers: machines are so logical, so orderly. If we didn't have people, everything would work so much better. Yup, that's how I used to think.

My work with that committee changed my view of design. Today, I realize that design presents a fascinating interplay of technology and psychology, that the designers must understand both. Engineers still tend to believe in logic. They often explain to me in great, logical detail, why their designs are good, powerful, and wonderful. "Why are people having problems?" they wonder. "You are being too logical," I say. "You are designing for people the way you would like them to be, not for the way they really are."

When the engineers object, I ask whether they have ever made an error, perhaps turning on or off the wrong light, or the wrong stove burner. "Oh yes," they say, "but those were errors." That's the point: even experts make errors. So we must design our machines on the assumption that people will make errors. (Chapter 5 provides a detailed analysis of human error.)

Human-Centered Design

People are frustrated with everyday things. From the ever-increasing complexity of the automobile dashboard, to the increasing automation in the home with its internal networks, complex music, video, and game systems for entertainment and communication, and the increasing automation in the kitchen, everyday life sometimes seems like a never-ending fight against confusion, continued errors, frustration, and a continual cycle of updating and maintaining our belongings.

In the multiple decades that have elapsed since the first edition of this book was published, design has gotten better. There are now many books and courses on the topic. But even though much has improved, the rapid rate of technology change outpaces the advances in design. New technologies, new applications, and new methods of interaction are continually arising and evolving. New industries spring up. Each new development seems to repeat the mistakes of the earlier ones; each new field requires time before it, too, adopts the principles of good design. And each new invention of technology or interaction technique requires experimentation and study before the principles of good design can be fully integrated into practice. So, yes, things are getting better, but as a result, the challenges are ever present.

The solution is human-centered design (HCD), an approach that puts human needs, capabilities, and behavior first, then designs to accommodate those needs, capabilities, and ways of behaving. Good design starts with an understanding of psychology and technology. Good design requires good communication, especially from machine to person, indicating what actions are possible, what is happening, and what is about to happen. Communication is especially important when things go wrong. It is relatively easy to design things that work smoothly and harmoniously as

TABLE 1.1. The Role of HCD and Design Specializations	
Experience design	These are areas of focus
Industrial design	
Interaction design	
Human-centered design	The process that ensures that the designs match the needs and capabilities of the people for whom they are intended

long as things go right. But as soon as there is a problem or a misunderstanding, the problems arise. This is where good design is essential. Designers need to focus their attention on the cases where things go wrong, not just on when things work as planned. Actually, this is where the most satisfaction can arise: when something goes wrong but the machine highlights the problems, then the person understands the issue, takes the proper actions, and the problem is solved. When this happens smoothly, the collaboration of person and device feels wonderful.

Human-centered design is a design philosophy. It means starting with a good understanding of people and the needs that the design is intended to meet. This understanding comes about primarily through observation, for people themselves are often unaware of their true needs, even unaware of the difficulties they are encountering. Getting the specification of the thing to be defined is one of the most difficult parts of the design, so much so that the HCD principle is to avoid specifying the problem as long as possible but instead to iterate upon repeated approximations. This is done through rapid tests of ideas, and after each test modifying the approach and the problem definition. The results can be products that truly meet the needs of people. Doing HCD within the rigid time, budget, and other constraints of industry can be a challenge: Chapter 6 examines these issues.

Where does HCD fit into the earlier discussion of the several different forms of design, especially the areas called industrial, interaction, and experience design? These are all compatible. HCD is a philosophy and a set of procedures, whereas the others are areas of focus (see Table 1.1). The philosophy and procedures of HCD add deep consideration and study of human needs to the design process, whatever the product or service, whatever the major focus.

Fundamental Principles of Interaction

Great designers produce pleasurable experiences. *Experience*: note the word. Engineers tend not to like it; it is too subjective. But when I ask them about their favorite automobile or test equipment, they will smile delightedly as they discuss the fit and finish, the sensation of power during acceleration, their ease of control while shifting or steering, or the wonderful feel of the knobs and switches on the instrument. Those are experiences.

Experience is critical, for it determines how fondly people remember their interactions. Was the overall experience positive, or was it frustrating and confusing? When our home technology behaves in an uninterpretable fashion we can become confused, frustrated, and even angry—all strong negative emotions. When there is understanding it can lead to a feeling of control, of mastery, and of satisfaction or even pride—all strong positive emotions. Cognition and emotion are tightly intertwined, which means that the designers must design with both in mind.

When we interact with a product, we need to figure out how to work it. This means discovering what it does, how it works, and what operations are possible: discoverability. Discoverability results from appropriate application of five fundamental psychological concepts covered in the next few chapters: *affordances, signifiers, constraints, mappings*, and *feedback*. But there is a sixth principle, perhaps most important of all: the *conceptual model* of the system. It is the conceptual model that provides true understanding. So I now turn to these fundamental principles, starting with affordances, signifiers, mappings, and feedback, then moving to conceptual models. Constraints are covered in Chapters 3 and 4.

AFFORDANCES

We live in a world filled with objects, many natural, the rest artificial. Every day we encounter thousands of objects, many of them new to us. Many of the new objects are similar to ones we already know, but many are unique, yet we manage quite well. How do we do this? Why is it that when we encounter many unusual natural objects, we know how to interact with them? Why is this true with many of the artificial, human-made objects we encounter? The answer lies with a few basic principles. Some of the most important of these principles come from a consideration of affordances.

The term *affordance* refers to the relationship between a physical object and a person (or for that matter, any interacting agent, whether animal or human, or even machines and robots). An affordance is a relationship between the properties of an object and the capabilities of the agent that determine just how the object could possibly be used. A chair affords ("is for") support and, therefore, affords sitting. Most chairs can also be carried by a single person (they afford lifting), but some can only be lifted by a strong person or by a team of people. If young or relatively weak people cannot lift a chair, then for these people, the chair does not have that affordance, it does not afford lifting.

The presence of an affordance is jointly determined by the qualities of the object and the abilities of the agent that is interacting. This relational definition of affordance gives considerable difficulty to many people. We are used to thinking that properties are associated with objects. But affordance is not a property. An affordance is a relationship. Whether an affordance exists depends upon the properties of both the object and the agent.

Glass affords transparency. At the same time, its physical structure blocks the passage of most physical objects. As a result, glass affords seeing through and support, but not the passage of air or most physical objects (atomic particles can pass through glass). The blockage of passage can be considered an anti-affordance—the prevention of interaction. To be effective, affordances and anti-affordances have to be discoverable—perceivable. This poses a difficulty with glass. The reason we like glass is its relative invisibility, but this aspect, so useful in the normal window, also hides its anti-affordance property of blocking passage. As a result, birds often try to fly through windows. And every year, numerous people injure themselves when they walk (or run) through closed glass doors or large picture windows. If an affordance or anti-affordance cannot be perceived, some means of signaling its presence is required: I call this property a *signifier* (discussed in the next section).

The notion of affordance and the insights it provides originated with J. J. Gibson, an eminent psychologist who provided many advances to our understanding of human perception. I had interacted with him over many years, sometimes in formal conferences and seminars, but most fruitfully over many bottles of beer, late at night, just talking. We disagreed about almost everything. I was an engineer who became a cognitive psychologist, trying to understand how the mind works. He started off as a Gestalt psychologist, but then developed an approach that is today named after him: Gibsonian psychology, an ecological approach to perception. He argued that the world contained the clues and that people simply picked them up through "direct perception." I argued that nothing could be direct: the brain had to process the information arriving at the sense organs to put together a coherent interpretation. "Nonsense," he loudly proclaimed; "it requires no interpretation: it is directly perceived." And then he would put his hand to his ears, and with a triumphant flourish, turn off his hearing aids: my counterarguments would fall upon deaf ears—literally.

When I pondered my question—how do people know how to act when confronted with a novel situation—I realized that a large part of the answer lay in Gibson's work. He pointed out that all the senses work together, that we pick up information about the world by the combined result of all of them. "Information pickup" was one of his favorite phrases, and Gibson believed that the combined information picked up by all of our sensory apparatus—sight, sound, smell, touch, balance, kinesthetic, acceleration, body position—determines our perceptions without the need for internal processing or cognition. Although he and I disagreed about the role played by the brain's internal processing, his brilliance was in focusing attention on the rich amount of information present in the world. Moreover, the physical objects conveyed important information about how people could interact with them, a property he named "affordance."

Affordances exist even if they are not visible. For designers, their visibility is critical: visible affordances provide strong clues to the operations of things. A flat plate mounted on a door affords pushing. Knobs afford turning, pushing, and pulling. Slots are for inserting things into. Balls are for throwing or bouncing. Perceived affordances help people figure out what actions are possible without the need for labels or instructions. I call the signaling component of affordances *signifiers.*

SIGNIFIERS

Are affordances important to designers? The first edition of this book introduced the term *affordances* to the world of design. The design community loved the concept and affordances soon propagated into the instruction and writing about design. I soon found mention of the term everywhere. Alas, the term became used in ways that had nothing to do with the original.

Many people find affordances difficult to understand because they are relationships, not properties. Designers deal with fixed properties, so there is a temptation to say that the property is an affordance. But that is not the only problem with the concept of affordances.

Designers have practical problems. They need to know how to design things to make them understandable. They soon discovered that when working with the graphical designs for electronic displays, they needed a way to designate which parts could be touched, slid upward, downward, or sideways, or tapped upon. The actions could be done with a mouse, stylus, or fingers. Some systems responded to body motions, gestures, and spoken words, with no touching of any physical device. How could designers describe what they were doing? There was no word that fit, so they took the closest existing word—*affordance.* Soon designers were saying such things as, "I put an affordance there," to describe why they displayed a circle on a screen to indicate where the person should touch, whether by mouse or by finger. "No," I said, "that is not an affordance. That is a way of communicating where the touch should be. You are communicating where to do the touching: the affordance of touching exists on the entire screen: you are trying to signify *where* the touch should take place. That's not the same thing as saying *what* action is possible."

Not only did my explanation fail to satisfy the design community, but I myself was unhappy. Eventually I gave up: designers needed a word to describe what they were doing, so they chose *affordance*. What alternative did they have? I decided to provide a better answer: *signifiers*. Affordances determine what actions are possible. Signifiers communicate where the action should take place. We need both.

People need some way of understanding the product or service they wish to use, some sign of what it is for, what is happening, and what the alternative actions are. People search for clues, for any sign that might help them cope and understand. It is the sign that is important, anything that might signify meaningful information. Designers need to provide these clues. What people need, and what designers must provide, are signifiers. Good design requires, among other things, good communication of the purpose, structure, and operation of the device to the people who use it. That is the role of the signifier.

The term *signifier* has had a long and illustrious career in the exotic field of semiotics, the study of signs and symbols. But just as I appropriated *affordance* to use in design in a manner somewhat different than its inventor had intended, I use *signifier* in a somewhat different way than it is used in semiotics. For me, the term *signifier* refers to any mark or sound, any perceivable indicator that communicates appropriate behavior to a person.

Signifiers can be deliberate and intentional, such as the sign PUSH on a door, but they may also be accidental and unintentional, such as our use of the visible trail made by previous people walking through a field or over a snow-covered terrain to determine the best path. Or how we might use the presence or absence of people waiting at a train station to determine whether we have missed the train. (I explain these ideas in more detail in my book *Living with Complexity*.)

A.

B. C.

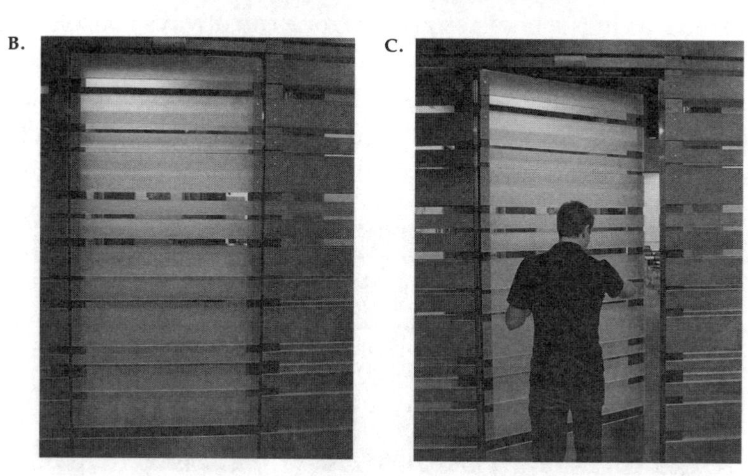

FIGURE 1.2. Problem Doors: Signifiers Are Needed. Door hardware can signal whether to push or pull without signs, but the hardware of the two doors in the upper photo, A, are identical even though one should be pushed, the other pulled. The flat, ribbed horizontal bar has the obvious perceived affordance of pushing, but as the signs indicate, the door on the left is to be pulled, the one on the right is to be pushed. In the bottom pair of photos, B and C, there are no visible signifiers or affordances. How does one know which side to push? Trial and error. When external signifiers—signs—have to be added to something as simple as a door, it indicates bad design. (Photographs by the author.)

The signifier is an important communication device to the recipient, whether or not communication was intended. It doesn't matter whether the useful signal was deliberately placed or whether it is incidental: there is no necessary distinction. Why should it matter whether a flag was placed as a deliberate clue to wind direction (as is done at airports or on the masts of sailboats) or was there as an advertisement or symbol of pride in one's country (as is done on public buildings). Once I interpret a flag's motion to indicate wind direction, it does not matter why it was placed there.

Consider a bookmark, a deliberately placed signifier of one's place in reading a book. But the physical nature of books also makes a bookmark an accidental signifier, for its placement also indicates how much of the book remains. Most readers have learned to use this accidental signifier to aid in their enjoyment of the reading. With few pages left, we know the end is near. And if the reading is torturous, as in a school assignment, one can always console oneself by knowing there are "only a few more pages to get through." Electronic book readers do not have the physical structure of paper books, so unless the software designer deliberately provides a clue, they do not convey any signal about the amount of text remaining.

A. B.

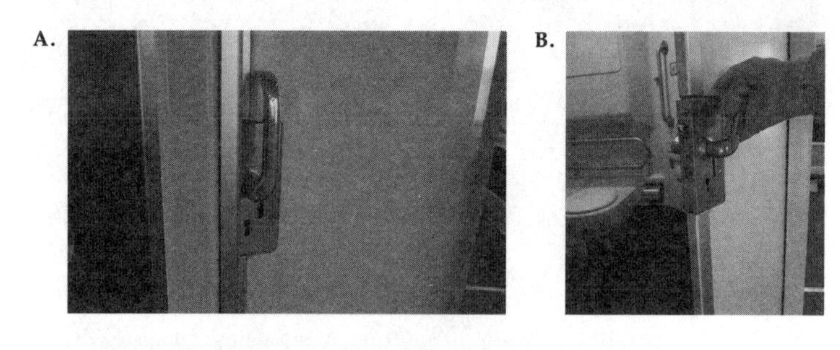

C.

FIGURE 1.3. **Sliding Doors: Seldom Done Well.** Sliding doors are seldom signified properly. The top two photographs show the sliding door to the toilet on an Amtrak train in the United States. The handle clearly signifies "pull," but in fact, it needs to be rotated and the door slid to the right. The owner of the store in Shanghai, China, Photo C, solved the problem with a sign. "DON'T PUSH!" it says, in both English and Chinese. Amtrak's toilet door could have used a similar kind of sign. (Photographs by the author.)

Whatever their nature, planned or accidental, signifiers provide valuable clues as to the nature of the world and of social activities. For us to function in this social, technological world, we need to develop internal models of what things mean, of how they operate. We seek all the clues we can find to help in this enterprise, and in this way, we are detectives, searching for whatever guidance we might find. If we are fortunate, thoughtful designers provide the clues for us. Otherwise, we must use our own creativity and imagination.

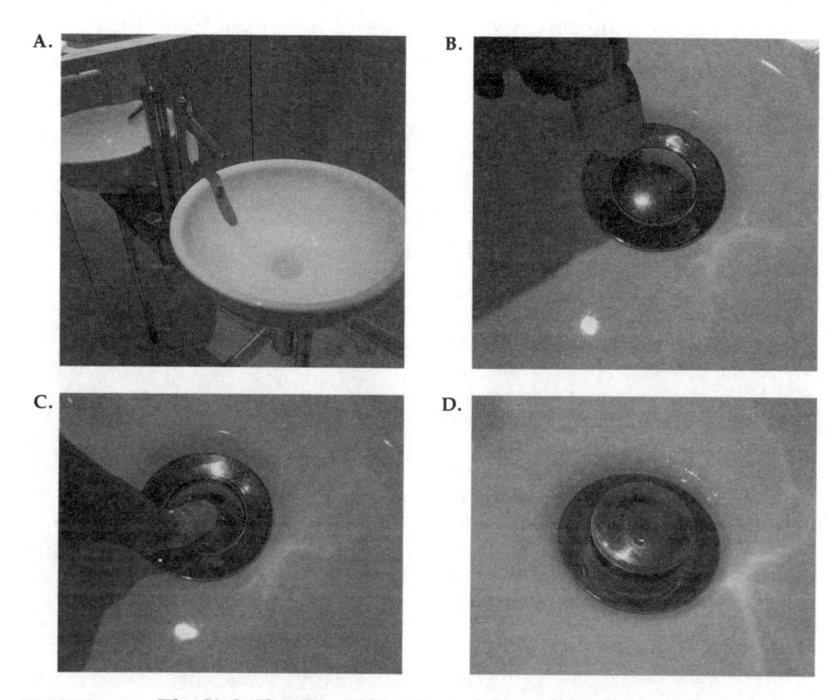

FIGURE 1.4. The Sink That Would Not Drain: Where Signifiers Fail. I washed my hands in my hotel sink in London, but then, as shown in Photo A, was left with the question of how to empty the sink of the dirty water. I searched all over for a control: none. I tried prying open the sink stopper with a spoon (Photo B): failure. I finally left my hotel room and went to the front desk to ask for instructions. (Yes, I actually did.) "Push down on the stopper," I was told. Yes, it worked (Photos C and D). But how was anyone to ever discover this? And why should I have to put my clean hands back into the dirty water to empty the sink? The problem here is not just the lack of signifier, it is the faulty decision to produce a stopper that requires people to dirty their clean hands to use it. (Photographs by the author.)

Affordances, perceived affordances, and signifiers have much in common, so let me pause to ensure that the distinctions are clear.

Affordances represent the possibilities in the world for how an agent (a person, animal, or machine) can interact with something. Some affordances are perceivable, others are invisible. Signifiers are signals. Some signifiers are signs, labels, and drawings placed in the world, such as the signs labeled "push," "pull," or "exit" on doors, or arrows and diagrams indicating what is to be acted upon or in which direction to gesture, or other instructions. Some signifiers are simply the perceived affordances, such as the handle of a door or the physical structure of a switch. Note that some perceived affordances may not be real: they may look like doors or places to push, or an impediment to entry, when in fact they are not. These are misleading signifiers, oftentimes accidental but sometimes purposeful, as when trying to keep people from doing actions for which they are not qualified, or in games, where one of the challenges is to figure out what is real and what is not.

FIGURE 1.5. Accidental Affordances Can Become Strong Signifiers. This wall, at the Industrial Design department of KAIST, in Korea, provides an anti-affordance, preventing people from falling down the stair shaft. Its top is flat, an accidental by-product of the design. But flat surfaces afford support, and as soon as one person discovers it can be used to dispose of empty drink containers, the discarded container becomes a signifier, telling others that it is permissible to discard their items there. (Photographs by the author.)

A.

B.

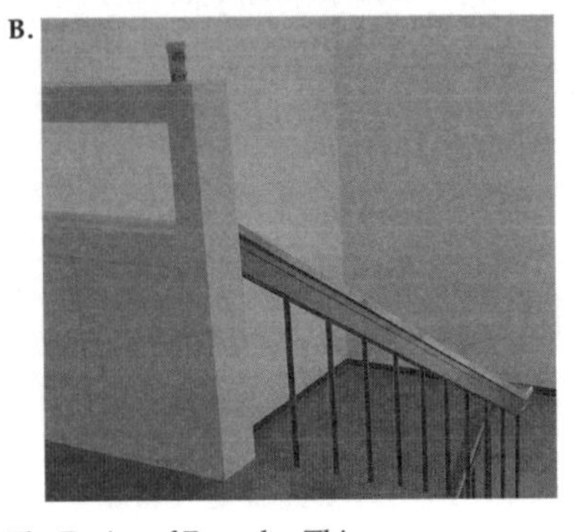

C.

My favorite example of a misleading signifier is a row of vertical pipes across a service road that I once saw in a public park. The pipes obviously blocked cars and trucks from driving on that road: they were good examples of anti-affordances. But to my great surprise, I saw a park vehicle simply go through the pipes. Huh? I walked over and examined them: the pipes were made of rubber, so vehicles could simply drive right over them. A very clever signifier, signaling a blocked road (via an apparent anti-affordance) to the average person, but permitting passage for those who knew.

To summarize:

- Affordances are the possible interactions between people and the environment. Some affordances are perceivable, others are not.
- Perceived affordances often act as signifiers, but they can be ambiguous.
- Signifiers signal things, in particular what actions are possible and how they should be done. Signifiers must be perceivable, else they fail to function.

In design, signifiers are more important than affordances, for they communicate how to use the design. A signifier can be words, a graphical illustration, or just a device whose perceived affordances are unambiguous. Creative designers incorporate the signifying part of the design into a cohesive experience. For the most part, designers can focus upon signifiers.

Because affordances and signifiers are fundamentally important principles of good design, they show up frequently in the pages of this book. Whenever you see hand-lettered signs pasted on doors, switches, or products, trying to explain how to work them, what to do and what not to do, you are also looking at poor design.

AFFORDANCES AND SIGNIFIERS: A CONVERSATION

A designer approaches his mentor. He is working on a system that recommends restaurants to people, based upon their preferences and those of their friends. But in his tests, he discovered that people never used all of the features. "Why not?" he asks his mentor.

(With apologies to Socrates.)

DESIGNER	MENTOR
I'm frustrated; people aren't using our application properly.	Can you tell me about it?
The screen shows the restaurant that we recommend. It matches their preferences, and their friends like it as well. If they want to see other recommendations, all they have to do is swipe left or right. To learn more about a place, just swipe up for a menu or down to see if any friends are there now. People seem to find the other recommendations, but not the menus or their friends? I don't understand.	Why do you think this might be?
I don't know. Should I add some affordances? Suppose I put an arrow on each edge and add a label saying what they do.	That is very nice. But why do you call these affordances? They could already do the actions. Weren't the affordances already there?
Yes, you have a point. But the affordances weren't visible. I made them visible.	Very true. You added a signal of what to do.
Yes, isn't that what I said?	Not quite—you called them affordances even though they afford nothing new: they signify what to do and where to do it. So call them by their right name: *"signifiers."*
Oh, I see. But then why do designers care about affordances? Perhaps we should focus our attention on signifiers.	You speak wisely. Communication is a key to good design. And a key to communication is the signifier.
Oh. Now I understand my confusion. Yes, a signifier is what signifies. It is a sign. Now it seems perfectly obvious.	Profound ideas are always obvious once they are understood.

MAPPING

Mapping is a technical term, borrowed from mathematics, meaning the relationship between the elements of two sets of things. Suppose there are many lights in the ceiling of a classroom or auditorium and a row of light switches on the wall at the front of the

FIGURE 1.6. Signifiers on a Touch Screen. The arrows and icons are signifiers: they provide signals about the permissible operations for this restaurant guide. Swiping left or right brings up new restaurant recommendations. Swiping up reveals the menu for the restaurant being displayed; swiping down, friends who recommend the restaurant.

room. The mapping of switches to lights specifies which switch controls which light.

Mapping is an important concept in the design and layout of controls and displays. When the mapping uses spatial correspondence between the layout of the controls and the devices being controlled, it is easy to determine how to use them. In steering a car, we rotate the steering wheel clockwise to cause the car to turn right: the top of the wheel moves in the same direction as the car. Note that other choices could have been made. In early cars, steering was controlled by a variety of devices, including tillers, handlebars, and reins. Today, some vehicles use joysticks, much as in a computer game. In cars that used tillers, steering was done much as one steers a boat: move the tiller to the left to turn to the right. Tractors, construction equipment such as bulldozers and cranes, and military tanks that have tracks instead of wheels use separate controls for the speed and direction of each track: to turn right, the left track is increased in speed, while the right track is slowed or even reversed. This is also how a wheelchair is steered.

All of these mappings for the control of vehicles work because each has a compelling conceptual model of how the operation of the control affects the vehicle. Thus, if we speed up the left wheel of a wheelchair while stopping the right wheel, it is easy to imagine the chair's pivoting on the right wheel, circling to the right. In a small boat, we can understand the tiller by realizing that pushing the tiller to the left causes the ship's rudder to move to the right and the resulting force of the water on the rudder slows down the right side of the boat, so that the boat rotates to the right. It doesn't matter whether these conceptual models are accurate: what matters is that they provide a clear way of remembering and understanding the mappings. The relationship between a control and its results is easiest to learn wherever there is an understandable mapping between the controls, the actions, and the intended result.

Natural mapping, by which I mean taking advantage of spatial analogies, leads to immediate understanding. For example, to move an object up, move the control up. To make it easy to determine which control works which light in a large room or auditorium, arrange the controls in the same pattern as the lights. Some natural mappings are cultural or biological, as in the universal standard that moving the hand up signifies more, moving it down signifies less, which is why it is appropriate to use vertical position to represent intensity or amount. Other natural mappings follow from the principles of perception and allow for the natural grouping or patterning of controls and feedback. Groupings and proximity are important principles from Gestalt psychology that can be used to map controls to function: related controls should be grouped together. Controls should be close to the item being controlled.

Note that there are many mappings that feel "natural" but in fact are specific to a particular culture: what is natural for one culture is not necessarily natural for another. In Chapter 3, I discuss how

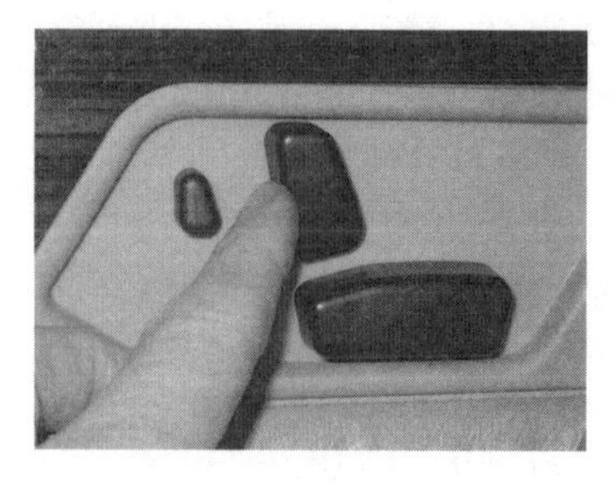

FIGURE 1.7. **Good Mapping: Automobile Seat Adjustment Control.** This is an excellent example of natural mapping. The control is in the shape of the seat itself: the mapping is straightforward. To move the front edge of the seat higher, lift up on the front part of the button. To make the seat back recline, move the button back. The same principle could be applied to much more common objects. This particular control is from Mercedes-Benz, but this form of mapping is now used by many automobile companies. (Photograph by the author.)

different cultures view time, which has important implications for some kinds of mappings.

A device is easy to use when the set of possible actions is visible, when the controls and displays exploit natural mappings. The principles are simple but rarely incorporated into design. Good design takes care, planning, thought, and an understanding of how people behave.

FEEDBACK

Ever watch people at an elevator repeatedly push the Up button, or repeatedly push the pedestrian button at a street crossing? Ever drive to a traffic intersection and wait an inordinate amount of time for the signals to change, wondering all the time whether the detection circuits noticed your vehicle (a common problem with bicycles)? What is missing in all these cases is feedback: some way of letting you know that the system is working on your request.

Feedback—communicating the results of an action—is a well-known concept from the science of control and information theory. Imagine trying to hit a target with a ball when you cannot see the target. Even as simple a task as picking up a glass with the hand requires feedback to aim the hand properly, to grasp the glass, and to lift it. A misplaced hand will spill the contents, too hard a grip will break the glass, and too weak a grip will allow it to fall. The human nervous system is equipped with numerous feedback mechanisms, including visual, auditory, and touch sensors, as well as vestibular and proprioceptive systems that monitor body position and muscle and limb movements. Given the importance of feedback, it is amazing how many products ignore it.

Feedback must be immediate: even a delay of a tenth of a second can be disconcerting. If the delay is too long, people often give up, going off to do other activities. This is annoying to the people, but it can also be wasteful of resources when the system spends considerable time and effort to satisfy the request, only to find that the intended recipient is no longer there. Feedback must also be informative. Many companies try to save money by using inexpensive lights or sound generators for feedback. These simple light flashes or beeps are usually more annoying than useful. They tell us that something has happened, but convey very little information about what has happened, and then nothing about what we should do about it. When the signal is auditory, in many cases we cannot even be certain which device has created the sound. If the signal is a light, we may miss it unless our eyes are on the correct spot at the correct time. Poor feedback can be worse than no feedback at all, because it is distracting, uninformative, and in many cases irritating and anxiety-provoking.

Too much feedback can be even more annoying than too little. My dishwasher likes to beep at three a.m. to tell me that the wash is done, defeating my goal of having it work in the middle of the night so as not to disturb anyone (and to use less expensive electricity). But worst of all is inappropriate, uninterpretable feedback. The irritation caused by a "backseat driver" is well enough known that it is the staple of numerous jokes. Backseat drivers are often correct, but their remarks and comments can be so numerous and continuous that instead of helping, they become an irritating distraction. Machines that give too much feedback are like backseat drivers. Not only is it distracting to be subjected to continual flashing lights, text announcements, spoken voices, or beeps and boops, but it can be dangerous. Too many announcements cause people to ignore all of them, or wherever possible, disable all of them, which means that critical and important ones are apt to be missed. Feedback is essential, but not when it gets in the way of other things, including a calm and relaxing environment.

Poor design of feedback can be the result of decisions aimed at reducing costs, even if they make life more difficult for people. Rather than use multiple signal lights, informative displays, or rich, musical sounds with varying patterns, the focus upon cost reduction forces the design to use a single light or sound to convey multiple types of information. If the choice is to use a light, then one flash might mean one thing; two rapid flashes, something else. A long flash might signal yet another state; and a long flash followed by a brief one, yet another. If the choice is to use a sound, quite often the least expensive sound device is selected, one that can only produce a high-frequency beep. Just as with the lights, the only way to signal different states of the machine is by beeping different patterns. What do all these different patterns mean? How can we possibly learn and remember them? It doesn't help that every different machine uses a different pattern of lights or beeps, sometimes with the same patterns meaning contradictory things for different machines. All the beeps sound alike, so it often isn't even possible to know which machine is talking to us.

Feedback has to be planned. All actions need to be confirmed, but in a manner that is unobtrusive. Feedback must also be prioritized, so that unimportant information is presented in an unobtrusive fashion, but important signals are presented in a way that does capture attention. When there are major emergencies, then even important signals have to be prioritized. When every device is signaling a major emergency, nothing is gained by the resulting cacophony. The continual beeps and alarms of equipment can be dangerous. In many emergencies, workers have to spend valuable time turning off all the alarms because the sounds interfere with the concentration required to solve the problem. Hospital operating rooms, emergency wards. Nuclear power control plants. Airplane cockpits. All can become confusing, irritating, and life-endangering places because of excessive feedback, excessive alarms, and incompatible message coding. Feedback is essential, but it has to be done correctly. Appropriately.

CONCEPTUAL MODELS

A conceptual model is an explanation, usually highly simplified, of how something works. It doesn't have to be complete or even accurate as long as it is useful. The files, folders, and icons you see displayed on a computer screen help people create the conceptual model of documents and folders inside the computer, or of apps or applications residing on the screen, waiting to be summoned. In fact, there are no folders inside the computer—those are effective conceptualizations designed to make them easier to use. Sometimes these depictions can add to the confusion, however. When reading e-mail or visiting a website, the material appears to be on the device, for that is where it is displayed and manipulated. But in fact, in many cases the actual material is "in the cloud," located on some distant machine. The conceptual model is of one, coherent image, whereas it may actually consist of parts, each located on different machines that could be almost anywhere in the world. This simplified model is helpful for normal usage, but if the network connection to the cloud services is interrupted, the result can be confusing. Information is still on their screen, but users can no longer save it or retrieve new things: their conceptual model offers no explanation. Simplified models are valuable only as long as the assumptions that support them hold true.

There are often multiple conceptual models of a product or device. People's conceptual models for the way that regenerative braking in a hybrid or electrically powered automobile works are quite different for average drivers than for technically sophisticated drivers, different again for whoever must service the system, and yet different again for those who designed the system.

Conceptual models found in technical manuals and books for technical use can be detailed and complex. The ones we are concerned with here are simpler: they reside in the minds of the people who are using the product, so they are also "mental models." Mental models, as the name implies, are the conceptual models in people's minds that represent their understanding of how things work. Different people may hold different mental models of the same item. Indeed, a single person might have multiple models of the same item, each dealing with a different aspect of its operation: the models can even be in conflict.

Conceptual models are often inferred from the device itself. Some models are passed on from person to person. Some come from manuals. Usually the device itself offers very little assistance, so the model is constructed by experience. Quite often these models are erroneous, and therefore lead to difficulties in using the device.

The major clues to how things work come from their perceived structure—in particular from signifiers, affordances, constraints, and mappings. Hand tools for the shop, gardening, and the house tend to make their critical parts sufficiently visible that concep-

FIGURE 1.8. Junghans Mega 1000 Digital Radio Controlled Watch. There is no good conceptual model for understanding the operation of my watch. It has five buttons with no hints as to what each one does. And yes, the buttons do different things in their different modes. But it is a very nice-looking watch, and always has the exact time because it checks official radio time stations. (The top row of the display is the date: Wednesday, February 20, the eighth week of the year.) (Photograph by the author.)

tual models of their operation and function are readily derived. Consider a pair of scissors: you can see that the number of possible actions is limited. The holes are clearly there to put something into, and the only logical things that will fit are fingers. The holes are both affordances—they allow the fingers to be inserted—and signifiers—they indicate where the fingers are to go. The sizes of the holes provide constraints to limit the possible fingers: a big hole suggests several fingers; a small hole, only one. The mapping between holes and fingers—the set of possible operations—is signified and constrained by the holes. Moreover, the operation is not sensitive to finger placement: if you use the wrong fingers (or the wrong hand), the scissors still work, although not as comfortably. You can figure out the scissors because their operating parts are visible and the implications clear. The conceptual model is obvious, and there is effective use of signifiers, affordances, and constraints.

What happens when the device does not suggest a good conceptual model? Consider my digital watch with five buttons: two along the top, two along the bottom, and one on the left side (Figure 1.8). What is each button for? How would you set the time? There is no way to tell—no evident relationship between the operating controls and the functions, no constraints, no apparent mappings. Moreover, the buttons have multiple ways of being used. Two of the buttons do different things when pushed quickly or when kept depressed for several seconds. Some operations require simultaneous depression of several of the buttons. The only way to tell how to work the watch is to read the manual, over and over again. With the scissors, moving the handle makes the blades move. The watch provides no visible relationship between the buttons and the possible actions, no discernible relationship between the actions and the end results. I really like the watch: too bad I can't remember all the functions.

Conceptual models are valuable in providing understanding, in predicting how things will behave, and in figuring out what to do when things do not go as planned. A good conceptual model allows us to predict the effects of our actions. Without a good model, we operate by rote, blindly; we do operations as we were told to do them; we can't fully appreciate why, what effects to expect, or what to do if things go wrong. As long as things work properly, we can manage. When things go wrong, however, or when we come upon a novel situation, then we need a deeper understanding, a good model.

For everyday things, conceptual models need not be very complex. After all, scissors, pens, and light switches are pretty simple devices. There is no need to understand the underlying physics or chemistry of each device we own, just the relationship between the controls and the outcomes. When the model presented to us is inadequate or wrong (or, worse, nonexistent), we can have difficulties. Let me tell you about my refrigerator.

I used to own an ordinary, two-compartment refrigerator—nothing very fancy about it. The problem was that I couldn't set the temperature properly. There were only two things to do: adjust the temperature of the freezer compartment and adjust the tempera-

FIGURE 1.9. Refrigerator Controls. Two compartments—fresh food and freezer—and two controls (in the fresh food unit). Your task: Suppose the freezer is too cold, the fresh food section just right. How would you adjust the controls so as to make the freezer warmer and keep the fresh food the same? (Photograph by the author.)

ture of the fresh food compartment. And there were two controls, one labeled "freezer," the other "refrigerator." What's the problem?

Oh, perhaps I'd better warn you. The two controls are not independent. The freezer control also affects the fresh food temperature, and the fresh food control also affects the freezer. Moreover, the manual warns that one should "always allow twenty-four (24) hours for the temperature to stabilize whether setting the controls for the first time or making an adjustment."

It was extremely difficult to regulate the temperature of my old refrigerator. Why? Because the controls suggest a false conceptual model. Two compartments, two controls, which implies that each control is responsible for the temperature of the compartment that carries its name: this conceptual model is shown in Figure 1.10A. It is wrong. In fact, there is only one thermostat and only one cooling mechanism. One control adjusts the thermostat setting, the other the relative proportion of cold air sent to each of the two compartments of the refrigerator. This is why the two controls interact: this conceptual model is shown in Figure 1.10B. In addition, there must be a temperature sensor, but there is no way of knowing where it is located. With the conceptual model suggested by the controls,

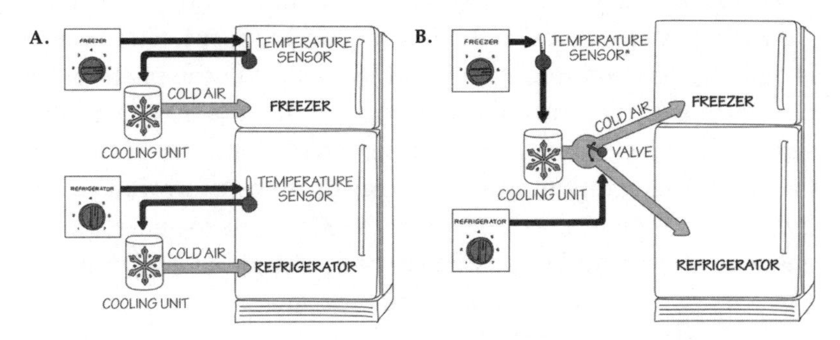

FIGURE 1.10. Two Conceptual Models for a Refrigerator. The conceptual model A is provided by the system image of the refrigerator as gleaned from the controls. Each control determines the temperature of the named part of the refrigerator. This means that each compartment has its own temperature sensor and cooling unit. This is wrong. The correct conceptual model is shown in B. There is no way of knowing where the temperature sensor is located so it is shown outside the refrigerator. The freezer control determines the freezer temperature (so is this where the sensor is located?). The refrigerator control determines how much of the cold air goes to the freezer and how much to the refrigerator.

adjusting the temperatures is almost impossible and always frustrating. Given the correct model, life would be much easier.

Why did the manufacturer suggest the wrong conceptual model? We will never know. In the twenty-five years since the publication of the first edition of this book, I have had many letters from people thanking me for explaining their confusing refrigerator, but never any communication from the manufacturer (General Electric). Perhaps the designers thought the correct model was too complex, that the model they were giving was easier to understand. But with the wrong conceptual model, it was impossible to set the controls. And even though I am convinced I knew the correct model, I still couldn't accurately adjust the temperatures because the refrigerator design made it impossible to discover which control was for the temperature sensor, which for the relative proportion of cold air, and in which compartment the sensor was located. The lack of immediate feedback for the actions did not help: it took twenty-four hours to see whether the new setting was appropriate. I shouldn't have to keep a laboratory notebook and do controlled experiments just to set the temperature of my refrigerator.

I am happy to say that I no longer own that refrigerator. Instead I have one that has two separate controls, one in the fresh food compartment, one in the freezer compartment. Each control is nicely calibrated in degrees and labeled with the name of the compartment it controls. The two compartments are independent: setting the temperature in one has no effect on the temperature in the other. This solution, although ideal, does cost more. But far less expensive solutions are possible. With today's inexpensive sensors and motors, it should be possible to have a single cooling unit with a motor-controlled valve controlling the relative proportion of cold air diverted to each compartment. A simple, inexpensive computer chip could regulate the cooling unit and valve position so that the temperatures in the two compartments match their targets. A bit more work for the engineering design team? Yes, but the results would be worth it. Alas, General Electric is still selling refrigerators with the very same controls and mechanisms that cause so much confusion. The photograph in Figure 1.9 is from a contemporary refrigerator, photographed in a store while preparing this book.

The System Image

People create mental models of themselves, others, the environment, and the things with which they interact. These are conceptual models formed through experience, training, and instruction. These models serve as guides to help achieve our goals and in understanding the world.

How do we form an appropriate conceptual model for the devices we interact with? We cannot talk to the designer, so we rely upon whatever information is available to us: what the device looks like, what we know from using similar things in the past, what was told to us in the sales literature, by salespeople and advertisements, by articles we may have read, by the product website and instruction manuals. I call the combined information available to us the *system image*. When the system image is incoherent or inappropriate, as in the case of the refrigerator, then the user cannot easily use the device. If it is incomplete or contradictory, there will be trouble.

As illustrated in Figure 1.11, the designer of the product and the person using the product form somewhat disconnected vertices of a triangle. The designer's conceptual model is the designer's conception of the product, occupying one vertex of the triangle. The product itself is no longer with the designer, so it is isolated as a second vertex, perhaps sitting on the user's kitchen counter. The system image is what can be perceived from the physical structure that has been built (including documentation, instructions, signifiers, and any information available from websites and help lines). The user's conceptual model comes from the system image, through interaction with the product, reading, searching for online information, and from whatever manuals are provided. The designer expects the user's model to be identical to the design model, but because designers cannot communicate directly with users, the entire burden of communication is on the system image.

FIGURE 1.11. The Designer's Model, the User's Model, and the System Image. The designer's conceptual model is the designer's conception of the look, feel, and operation of a product. The system image is what can be derived from the physical structure that has been built (including documentation). The user's mental model is developed through interaction with the product and the system image. Designers expect the user's model to be identical to their own, but because they cannot communicate directly with the user, the burden of communication is with the system image.

Figure 1.11 indicates why communication is such an important aspect of good design. No matter how brilliant the product, if people cannot use it, it will receive poor reviews. It is up to the designer to provide the appropriate information to make the product understandable and usable. Most important is the provision of a good conceptual model that guides the user when thing go wrong. With a good conceptual model, people can figure out what has happened and correct the things that went wrong. Without a good model, they struggle, often making matters worse.

Good conceptual models are the key to understandable, enjoyable products: good communication is the key to good conceptual models.

The Paradox of Technology

Technology offers the potential to make life easier and more enjoyable; each new technology provides increased benefits. At the same time, added complexities increase our difficulty and frustration with technology. The design problem posed by technological advances is enormous. Consider the wristwatch. A few decades ago, watches were simple. All you had to do was set the time and keep the watch wound. The standard control was the stem: a knob at the side of the watch. Turning the knob would wind the spring that provided power to the watch movement. Pulling out the knob and turning it rotated the hands. The operations were easy to learn and easy to do. There was a reasonable relationship between the turning of the knob and the resulting turning of the hands. The design even took into account human error. In its normal position, turning the stem wound the mainspring of the clock. The stem had to be pulled before it would engage the gears for setting the time. Accidental turns of the stem did no harm.

Watches in olden times were expensive instruments, manufactured by hand. They were sold in jewelry stores. Over time, with the introduction of digital technology, the cost of watches decreased rapidly, while their accuracy and reliability increased. Watches became tools, available in a wide variety of styles and shapes and with an ever-increasing number of functions. Watches were sold everywhere, from local shops to sporting goods stores to electronic stores. Moreover, accurate clocks were incorporated in many appliances, from phones to musical keyboards: many people no longer felt the need to wear a watch. Watches became inexpensive enough that the average person could own multiple watches. They became fashion accessories, where one changed the watch with each change in activity and each change of clothes.

In the modern digital watch, instead of winding the spring, we change the battery, or in the case of a solar-powered watch, ensure that it gets its weekly dose of light. The technology has allowed more functions: the watch can give the day of the week, the month, and the year; it can act as a stopwatch (which itself has several functions), a countdown timer, and an alarm clock (or two); it has the ability to show the time for different time zones; it can act as a counter and even as a calculator. My watch, shown in Figure 1.8, has many functions. It even has a radio receiver to allow it to set its time with official time stations around the world. Even so, it is far less complex than many that are available. Some watches have built-in compasses and barometers, accelerometers, and temperature gauges. Some have GPS and Internet receivers so they can display the weather and news, e-mail messages, and the latest from social networks. Some have built-in cameras. Some work with buttons, knobs, motion, or speech. Some detect gestures. The watch is no longer just an instrument for telling time: it has become a platform for enhancing multiple activities and lifestyles.

The added functions cause problems: How can all these functions fit into a small, wearable size? There are no easy answers. Many people have solved the problem by not using a watch. They use their phone instead. A cell phone performs all the functions much better than the tiny watch, while also displaying the time.

Now imagine a future where instead of the phone replacing the watch, the two will merge, perhaps worn on the wrist, perhaps on the head like glasses, complete with display screen. The phone, watch, and components of a computer will all form one unit. We will have flexible displays that show only a tiny amount of information in their normal state, but that can unroll to considerable size. Projectors will be so small and light that they can be built into watches or phones (or perhaps rings and other jewelry), projecting their images onto any convenient surface. Or perhaps our devices won't have displays, but will quietly whisper the results into our ears, or simply use whatever display happens to be available: the display in the seatback of cars or airplanes, hotel room televisions, whatever is nearby. The devices will be able to do many useful things, but I fear they will also frustrate: so many things to control, so little space for controls or signifiers. The obvious solution is to use exotic gestures or spoken commands, but how will we learn, and then remember, them? As I discuss later, the best solution is for there to be agreed upon standards, so we need learn the controls only once. But as I also discuss, agreeing upon these is a complex process, with many competing forces hindering rapid resolution. We will see.

The same technology that simplifies life by providing more functions in each device also complicates life by making the device harder to learn, harder to use. This is the paradox of technology and the challenge for the designer.

The Design Challenge

Design requires the cooperative efforts of multiple disciplines. The number of different disciplines required to produce a successful product is staggering. Great design requires great designers, but that isn't enough: it also requires great management, because the hardest part of producing a product is coordinating all the many, separate disciplines, each with different goals and priorities. Each discipline has a different perspective of the relative importance of the many factors that make up a product. One discipline argues that it must be usable and understandable, another that it must be attractive, yet another that it has to be affordable. Moreover, the device has to be reliable, be able to be manufactured and serviced. It must be distinguishable from competing products and superior in critical dimensions such as price, reliability, appearance, and the functions it provides. Finally, people have to actually purchase it. It doesn't matter how good a product is if, in the end, nobody uses it.

Quite often each discipline believes its distinct contribution to be most important: "Price," argues the marketing representative, "price plus these features." "Reliable," insist the engineers. "We have to be able to manufacture it in our existing plants," say the manufacturing representatives. "We keep getting service calls," say the support people; "we need to solve those problems in the design." "You can't put all that together and still have a reasonable product," says the design team. Who is right? Everyone is right. The successful product has to satisfy all these requirements.

The hard part is to convince people to understand the viewpoints of the others, to abandon their disciplinary viewpoint and to think of the design from the viewpoints of the person who buys the product and those who use it, often different people. The viewpoint of the business is also important, because it does not matter how wonderful the product is if not enough people buy it. If a product does not sell, the company must often stop producing it, even if it is a great product. Few companies can sustain the huge cost of keeping an unprofitable product alive long enough for its sales to reach profitability—with new products, this period is usually measured in years, and sometimes, as with the adoption of high-definition television, decades.

Designing well is not easy. The manufacturer wants something that can be produced economically. The store wants something that will be attractive to its customers. The purchaser has several demands. In the store, the purchaser focuses on price and appearance, and perhaps on prestige value. At home, the same person will pay more attention to functionality and usability. The repair service cares about maintainability: how easy is the device to take apart, diagnose, and service? The needs of those concerned are different and often conflict. Nonetheless, if the design team has representatives from all the constituencies present at the same time, it is often possible to reach satisfactory solutions for all the needs. It is when the disciplines operate independently of one another that major clashes and deficiencies occur. The challenge is to use the principles of human-centered design to produce positive results, products that enhance lives and add to our pleasure and enjoyment. The goal is to produce a great product, one that is successful, and that customers love. It can be done.

CHAPTER TWO

THE PSYCHOLOGY OF EVERYDAY ACTIONS

During my family's stay in England, we rented a furnished house while the owners were away. One day, our landlady returned to the house to get some personal papers. She walked over to the old, metal filing cabinet and attempted to open the top drawer. It wouldn't open. She pushed it forward and backward, right and left, up and down, without success. I offered to help. I wiggled the drawer. Then I twisted the front panel, pushed down hard, and banged the front with the palm of one hand. The cabinet drawer slid open. "Oh," she said, "I'm sorry. I am so bad at mechanical things." No, she had it backward. It is the mechanical thing that should be apologizing, perhaps saying, "I'm sorry. I am so bad with people."

My landlady had two problems. First, although she had a clear goal (retrieve some personal papers) and even a plan for achieving that goal (open the top drawer of the filing cabinet, where those papers are kept), once that plan failed, she had no idea of what to do. But she also had a second problem: she thought the problem lay in her own lack of ability: she blamed herself, falsely.

How was I able to help? First, I refused to accept the false accusation that it was the fault of the landlady: to me, it was clearly a fault in the mechanics of the old filing cabinet that prevented the drawer from opening. Second, I had a conceptual model of how the cabinet worked, with an internal mechanism that held the door shut in normal usage, and the belief that the drawer mechanism was probably out of alignment. This conceptual model gave me a plan: wiggle the drawer. That failed. That caused me to modify my plan: wiggling may have been appropriate but not forceful enough, so I resorted to brute force to try to twist the cabinet back into its proper alignment. This felt good to me—the cabinet drawer moved slightly—but it still didn't open. So I resorted to the most powerful tool employed by experts the world around—I banged on the cabinet. And yes, it opened. In my mind, I decided (without any evidence) that my hit had jarred the mechanism sufficiently to allow the drawer to open.

This example highlights the themes of this chapter. First, how do people do things? It is easy to learn a few basic steps to perform operations with our technologies (and yes, even filing cabinets are technology). But what happens when things go wrong? How do we detect that they aren't working, and then how do we know what to do? To help understand this, I first delve into human psychology and a simple conceptual model of how people select and then evaluate their actions. This leads the discussion to the role of understanding (via a conceptual model) and of emotions: pleasure when things work smoothly and frustration when our plans are thwarted. Finally, I conclude with a summary of how the lessons of this chapter translate into principles of design.

How People Do Things: The Gulfs of Execution and Evaluation

When people use something, they face two gulfs: the Gulf of Execution, where they try to figure out how it operates, and the Gulf of Evaluation, where they try to figure out what happened (Figure 2.1). The role of the designer is to help people bridge the two gulfs.

In the case of the filing cabinet, there were visible elements that helped bridge the Gulf of Execution when everything was working perfectly. The drawer handle clearly signified that it should be pulled and the slider on the handle indicated how to release the catch that normally held the drawer in place. But when these operations failed, there then loomed a big gulf: what other operations could be done to open the drawer?

The Gulf of Evaluation was easily bridged, at first. That is, the catch was released, the drawer handle pulled, yet nothing happened. The lack of action signified a failure to reach the goal. But when other operations were tried, such as my twisting and pulling, the filing cabinet provided no more information about whether I was getting closer to the goal.

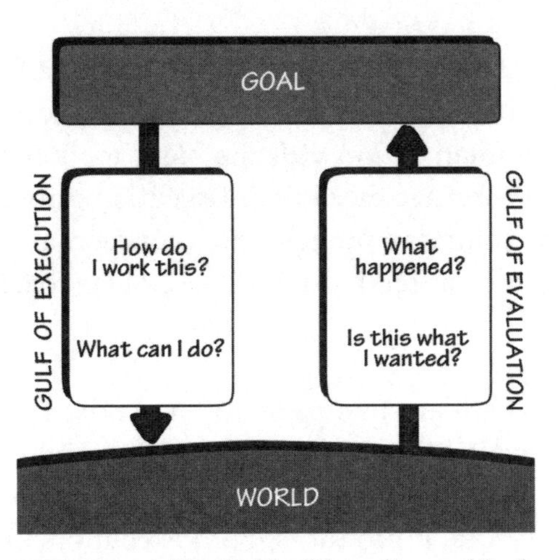

FIGURE 2.1. The Gulfs of Execution and Evaluation. When people encounter a device, they face two gulfs: the Gulf of Execution, where they try to figure out how to use it, and the Gulf of Evaluation, where they try to figure out what state it is in and whether their actions got them to their goal.

The Gulf of Evaluation reflects the amount of effort that the person must make to interpret the physical state of the device and to determine how well the expectations and intentions have been met. The gulf is small when the device provides information about its state in a form that is easy to get, is easy to interpret, and matches the way the person thinks about the system. What are the major design elements that help bridge the Gulf of Evaluation? Feedback and a good conceptual model.

The gulfs are present for many devices. Interestingly, many people do experience difficulties, but explain them away by blaming themselves. In the case of things they believe they should be capable of using—water faucets, refrigerator temperature controls, stove tops—they simply think, "I'm being stupid." Alternatively, for complicated-looking devices—sewing machines, washing machines, digital watches, or almost any digital controls—they simply give up, deciding that they are incapable of understanding them. Both explanations are wrong. These are the things of everyday household use. None of them has a complex underlying structure. The difficulties reside in their design, not in the people attempting to use them.

How can the designer help bridge the two gulfs? To answer that question, we need to delve more deeply into the psychology of human action. But the basic tools have already been discussed: We bridge the Gulf of Execution through the use of signifiers, constraints, mappings, and a conceptual model. We bridge the Gulf of Evaluation through the use of feedback and a conceptual model.

The Seven Stages of Action

There are two parts to an action: executing the action and then evaluating the results: doing and interpreting. Both execution and evaluation require understanding: how the item works and what results it produces. Both execution and evaluation can affect our emotional state.

Suppose I am sitting in my armchair, reading a book. It is dusk, and the light is getting dimmer and dimmer. My current activity is reading, but that goal is starting to fail because of the decreasing illumination. This realization triggers a new goal: get more light. How do I do that? I have many choices. I could open the curtains, move so that I sit where there is more light, or perhaps turn on a nearby light. This is the planning stage, determining which of the many possible plans of action to follow. But even when I decide to turn on the nearby light, I still have to determine how to get it done. I could ask someone to do it for me, I could use my left hand or my right. Even after I have decided upon a plan, I still have to specify how I will do it. Finally, I must execute—do—the action. When I am doing a frequent act, one for which I am quite experienced and skilled, most of these stages are subconscious. When I am still learning how to do it, determining the plan, specifying the sequence, and interpreting the result are conscious.

Suppose I am driving in my car and my action plan requires me to make a left turn at a street intersection. If I am a skilled driver, I don't have to give much conscious attention to specify or perform the action sequence. I think "left" and smoothly execute the required action sequence. But if I am just learning to drive, I have to think about each separate component of the action. I must apply the brakes and check for cars behind and around me, cars and pedestrians in front of me, and whether there are traffic signs or signals that I have to obey. I must move my feet back and forth between pedals and my hands to the turn signals and back to the steering wheel (while I try to remember just how my instructor told me I should position my hands while making a turn), and my visual attention is divided among all the activity around me, sometimes looking directly, sometimes rotating my head, and sometimes using the rear- and side-view mirrors. To the skilled driver, it is all easy and straightforward. To the beginning driver, the task seems impossible.

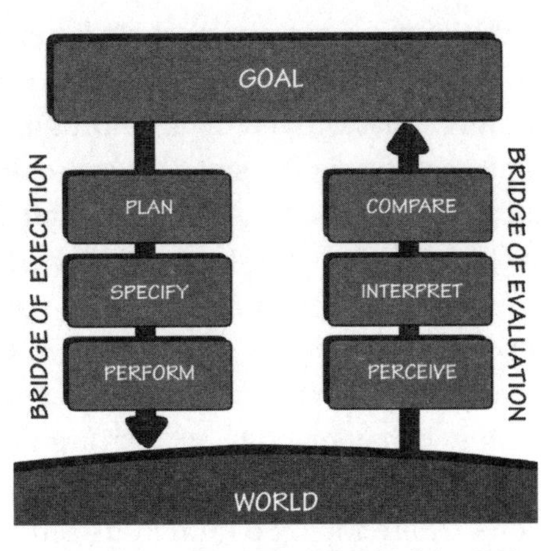

FIGURE 2.2. **The Seven Stages of the Action Cycle.** Putting all the stages together yields the three stages of execution (plan, specify, and perform), three stages of evaluation (perceive, interpret, and compare), and, of course, the goal: seven stages in all.

The specific actions bridge the gap between what we would like to have done (our goals) and all possible physical actions to achieve those goals. After we specify what actions to make, we must actually do them—the stages of execution. There are three stages of execution that follow from the goal: plan, specify, and perform (the left side of Figure 2.2). Evaluating what happened has three stages: first, perceiving what happened in the world; second, trying to make sense of it (interpreting it); and, finally, comparing what happened with what was wanted (the right side of Figure 2.2).

There we have it. Seven stages of action: one for goals, three for execution, and three for evaluation (Figure 2.2).

1. **Goal** (form the goal)
2. **Plan** (the action)
3. **Specify** (an action sequence)
4. **Perform** (the action sequence)
5. **Perceive** (the state of the world)
6. **Interpret** (the perception)
7. **Compare** (the outcome with the goal)

The seven-stage action cycle is simplified, but it provides a useful framework for understanding human action and for guiding design. It has proven to be helpful in designing interaction. Not all of the activity in the stages is conscious. Goals tend to be, but even they may be subconscious. We can do many actions, repeatedly cycling through the stages while being blissfully unaware that we are doing so. It is only when we come across something new or reach some impasse, some problem that disrupts the normal flow of activity, that conscious attention is required.

Most behavior does not require going through all stages in sequence; however, most activities will not be satisfied by single actions. There must be numerous sequences, and the whole activity may last hours or even days. There are multiple feedback loops in which the results of one activity are used to direct further ones, in which goals lead to subgoals, and plans lead to subplans. There are activities in which goals are forgotten, discarded, or reformulated.

Let's go back to my act of turning on the light. This is a case of event-driven behavior: the sequence starts with the world, causing evaluation of the state and the formulation of a goal. The trigger was an environmental event: the lack of light, which made reading difficult. This led to a violation of the goal of reading, so it led to a subgoal—get more light. But reading was not the high-level goal. For each goal, one has to ask, "Why is that the goal?" Why was I reading? I was trying to prepare a meal using a new recipe, so I needed to reread it before I started. Reading was thus a subgoal. But cooking was itself a subgoal. I was cooking in order to eat, which had the goal of satisfying my hunger. So the hierarchy of goals is roughly: satisfy hunger; eat; cook; read cookbook; get more light. This is called a root cause analysis: asking "Why?" until the ultimate, fundamental cause of the activity is reached.

The action cycle can start from the top, by establishing a new goal, in which case we call it goal-driven behavior. In this situation, the cycle starts with the goal and then goes through the three stages of execution. But the action cycle can also start from the bottom, triggered by some event in the world, in which case we call it either data-driven or event-driven behavior. In this situation, the cycle starts with the environment, the world, and then goes through the three stages of evaluation.

For many everyday tasks, goals and intentions are not well specified: they are opportunistic rather than planned. Opportunistic actions are those in which the behavior takes advantage of circumstances. Rather than engage in extensive planning and analysis, we go about the day's activities and do things as opportunities arise. Thus, we may not have planned to try a new café or to ask a question of a friend. Rather, we go through the day's activities, and if we find ourselves near the café or encountering the friend, then we allow the opportunity to trigger the appropriate activity. Otherwise, we might never get to that café or ask our friend the question. For crucial tasks we make special efforts to ensure that they get done. Opportunistic actions are less precise and certain than specified goals and intentions, but they result in less mental effort, less inconvenience, and perhaps more interest. Some of us adjust our lives around the expectation of opportunities. And sometimes, even for goal-driven behavior, we try to create world events that will ensure that the sequence gets completed. For example, sometimes when I must do an important task, I ask someone to set a deadline for me. I use the approach of that deadline to trigger the work. It may only be a few hours before the deadline that I actually get to work and do the job, but the important point is that it does get done. This self-triggering of external drivers is fully compatible with the seven-stage analysis.

The seven stages provide a guideline for developing new products or services. The gulfs are obvious places to start, for either gulf, whether of execution or evaluation, is an opportunity for product enhancement. The trick is to develop observational skills to detect them. Most innovation is done as an incremental enhancement of existing products. What about radical ideas, ones that introduce new product categories to the marketplace? These come about by reconsidering the goals, and always asking what the real goal is: what is called the *root cause* analysis.

Harvard Business School marketing professor Theodore Levitt once pointed out, "People don't want to buy a quarter-inch drill.

They want a quarter-inch hole!" Levitt's example of the drill implying that the goal is really a hole is only partially correct, however. When people go to a store to buy a drill, that is not their real goal. But why would anyone want a quarter-inch hole? Clearly that is an intermediate goal. Perhaps they wanted to hang shelves on the wall. Levitt stopped too soon.

Once you realize that they don't really want the drill, you realize that perhaps they don't really want the hole, either: they want to install their bookshelves. Why not develop methods that don't require holes? Or perhaps books that don't require bookshelves. (Yes, I know: electronic books, e-books.)

Human Thought: Mostly Subconscious

Why do we need to know about the human mind? Because things are designed to be used by people, and without a deep understanding of people, the designs are apt to be faulty, difficult to use, difficult to understand. That is why it is useful to consider the seven stages of action. The mind is more difficult to comprehend than actions. Most of us start by believing we already understand both human behavior and the human mind. After all, we are all human: we have all lived with ourselves all of our lives, and we like to think we understand ourselves. But the truth is, we don't. Most of human behavior is a result of subconscious processes. We are unaware of them. As a result, many of our beliefs about how people behave—including beliefs about ourselves—are wrong. That is why we have the multiple social and behavioral sciences, with a good dash of mathematics, economics, computer science, information science, and neuroscience.

Consider the following simple experiment. Do all three steps:

1. Wiggle the second finger of your hand.
2. Wiggle the third finger of the same hand.
3. Describe what you did differently those two times.

On the surface, the answer seems simple: I thought about moving my fingers and they moved. The difference is that I thought about a different finger each time. Yes, that's true. But how did that thought get transmitted into action, into the commands that caused different muscles in the arm to control the tendons that wiggled the fingers? This is completely hidden from consciousness.

The human mind is immensely complex, having evolved over a long period with many specialized structures. The study of the mind is the subject of multiple disciplines, including the behavioral and social sciences, cognitive science, neuroscience, philosophy, and the information and computer sciences. Despite many advances in our understanding, much still remains mysterious, yet to be learned. One of the mysteries concerns the nature of and distinction between those activities that are conscious and those that are not. Most of the brain's operations are subconscious, hidden beneath our awareness. It is only the highest level, what I call *reflective,* that is conscious.

Conscious attention is necessary to learn most things, but after the initial learning, continued practice and study, sometimes for thousands of hours over a period of years, produces what psychologists call "overlearning," Once skills have been overlearned, performance appears to be effortless, done automatically, with little or no awareness. For example, answer these questions:

What is the phone number of a friend?
What is Beethoven's phone number?
What is the capital of:

- Brazil?
- Wales?
- The United States?
- Estonia?

Think about how you answered these questions. The answers you knew come immediately to mind, but with no awareness of how that happened. You simply "know" the answer. Even the ones you got wrong came to mind without any awareness. You might have been aware of some doubt, but not of how the name entered your consciousness. As for the countries for which you didn't know the answer, you probably knew you didn't know those immediately, without effort. Even if you knew you knew, but couldn't quite recall it, you didn't know how you knew that, or what was happening as you tried to remember.

You might have had trouble with the phone number of a friend because most of us have turned over to our technology the job of remembering phone numbers. I don't know anybody's phone number—I barely remember my own. When I wish to call someone, I just do a quick search in my contact list and have the telephone place the call. Or I just push the "2" button on the phone for a few seconds, which autodials my home. Or in my auto, I can simply speak: "Call home." What's the number? I don't know: my technology knows. Do we count our technology as an extension of our memory systems? Of our thought processes? Of our mind?

What about Beethoven's phone number? If I asked my computer, it would take a long time, because it would have to search all the people I know to see whether any one of them was Beethoven. But you immediately discarded the question as nonsensical. You don't personally know Beethoven. And anyway, he is dead. Besides, he died in the early 1800s and the phone wasn't invented until the late 1800s. How do we know what we do not know so rapidly? Yet some things that we do know can take a long time to retrieve. For example, answer this:

> *In the house you lived in three houses ago, as you entered the front door, was the doorknob on the left or right?*

Now you have to engage in conscious, reflective problem solving, first to retrieve just which house is being talked about, and then what the correct answer is. Most people can determine the house, but have difficulty answering the question because they can readily imagine the doorknob on both sides of the door. The way to solve this problem is to imagine doing some activity, such as walking up to the front door while carrying heavy packages with both hands: how do you open the door? Alternatively, visualize yourself inside the house, rushing to the front door to open it for a visitor.

Usually one of these imagined scenarios provides the answer. But note how different the memory retrieval for this question was from the retrieval for the others. All these questions involved long-term memory, but in very different ways. The earlier questions were memory for factual information, what is called *declarative memory.* The last question could have been answered factually, but is usually most easily answered by recalling the activities performed to open the door. This is called *procedural memory.* I return to a discussion of human memory in Chapter 3.

Walking, talking, reading. Riding a bicycle or driving a car. Singing. All of these skills take considerable time and practice to master, but once mastered, they are often done quite automatically. For experts, only especially difficult or unexpected situations require conscious attention.

Because we are only aware of the reflective level of conscious processing, we tend to believe that all human thought is conscious. But it isn't. We also tend to believe that thought can be separated from emotion. This is also false. Cognition and emotion cannot be separated. Cognitive thoughts lead to emotions: emotions drive cognitive thoughts. The brain is structured to act upon the world, and every action carries with it expectations, and these expectations drive emotions. That is why much of language is based on physical metaphors, why the body and its interaction with the environment are essential components of human thought.

Emotion is highly underrated. In fact, the emotional system is a powerful information processing system that works in tandem with cognition. Cognition attempts to make sense of the world: emotion assigns value. It is the emotional system that determines whether a situation is safe or threatening, whether something that is happening is good or bad, desirable or not. Cognition provides understanding: emotion provides value judgments. A human without a working emotional system has difficulty making choices. A human without a cognitive system is dysfunctional.

Because much human behavior is subconscious—that is, it occurs without conscious awareness—we often don't know what we are about to do, say, or think until after we have done it. It's as if we had two minds: the subconscious and the conscious, which don't always talk to each other. Not what you have been taught? True, nonetheless. More and more evidence is accumulating that we use logic and reason after the fact, to justify our decisions to ourselves (to our conscious minds) and to others. Bizarre? Yes, but don't protest: enjoy it.

Subconscious thought matches patterns, finding the best possible match of one's past experience to the current one. It proceeds rapidly and automatically, without effort. Subconscious processing is one of our strengths. It is good at detecting general trends, at recognizing the relationship between what we now experience and what has happened in the past. And it is good at generalizing, at making predictions about the general trend, based on few examples. But subconscious thought can find matches that are inappropriate or wrong, and it may not distinguish the common from the rare. Subconscious thought is biased toward regularity and structure, and it is limited in formal power. It may not be capable of symbolic manipulation, of careful reasoning through a sequence of steps.

Conscious thought is quite different. It is slow and labored. Here is where we slowly ponder decisions, think through alternatives, compare different choices. Conscious thought considers first this approach, then that—comparing, rationalizing, finding explanations. Formal logic, mathematics, decision theory: these are the tools of conscious thought. Both conscious and subconscious modes of thought are powerful and essential aspects of human life. Both can provide insightful leaps and creative moments. And both are subject to errors, misconceptions, and failures.

Emotion interacts with cognition biochemically, bathing the brain with hormones, transmitted either through the bloodstream or through ducts in the brain, modifying the behavior of brain cells. Hormones exert powerful biases on brain operation. Thus, in tense, threatening situations, the emotional system triggers the release of hormones that bias the brain to focus upon relevant parts of the environment. The muscles tense in preparation for action. In calm, nonthreatening situations, the emotional system triggers the release of hormones that relax the muscles and bias the brain toward explo-

TABLE 2.1. Subconscious and Conscious Systems of Cognition	
Subconscious	**Conscious**
Fast	Slow
Automatic	Controlled
Multiple resources	Limited resources
Controls skilled behavior	Invoked for novel situations: when learning, when in danger, when things go wrong

ration and creativity. Now the brain is more apt to notice changes in the environment, to be distracted by events, and to piece together events and knowledge that might have seemed unrelated earlier.

A positive emotional state is ideal for creative thought, but it is not very well suited for getting things done. Too much, and we call the person scatterbrained, flitting from one topic to another, unable to finish one thought before another comes to mind. A brain in a negative emotional state provides focus: precisely what is needed to maintain attention on a task and finish it. Too much, however, and we get tunnel vision, where people are unable to look beyond their narrow point of view. Both the positive, relaxed state and the anxious, negative, and tense state are valuable and powerful tools for human creativity and action. The extremes of both states, however, can be dangerous.

Human Cognition and Emotion

The mind and brain are complex entities, still the topic of considerable scientific research. One valuable explanation of the levels of processing within the brain, applicable to both cognitive and emotional processing, is to think of three different levels of processing, each quite different from the other, but all working together in concert. Although this is a gross oversimplification of the actual processing, it is a good enough approximation to provide guidance in understanding human behavior. The approach I use here comes from my book *Emotional Design*. There, I suggested that a useful approximate model of human cognition and emotion is to consider three levels of processing: visceral, behavioral, and reflective.

THE VISCERAL LEVEL

The most basic level of processing is called *visceral.* This is sometimes referred to as "the lizard brain." All people have the same basic visceral responses. These are part of the basic protective mechanisms of the human affective system, making quick judgments about the environment: good or bad, safe or dangerous. The visceral system allows us to respond quickly and subconsciously, without conscious awareness or control. The basic biology of the visceral system minimizes its ability to learn. Visceral learning takes place primarily by sensitization or desensitization through such mechanisms as adaptation and classical conditioning. Visceral responses are fast and automatic. They give rise to the startle reflex for novel, unexpected events; for such genetically programmed behavior as fear of heights, dislike of the dark or very noisy environments, dislike of bitter tastes and the liking of sweet tastes, and so on. Note that the visceral level responds to the immediate present and produces an affective state, relatively unaffected by context or history. It simply assesses the situation: no cause is assigned, no blame, and no credit.

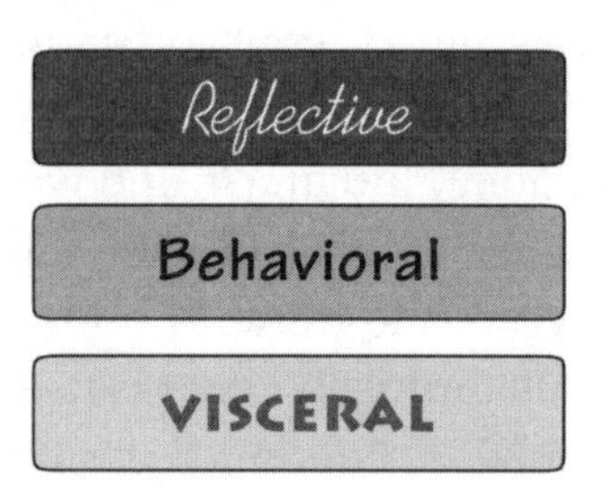

FIGURE 2.3. Three Levels of Processing: Visceral, Behavioral, and Reflective. Visceral and behavioral levels are subconscious and the home of basic emotions. The reflective level is where conscious thought and decision-making reside, as well as the highest level of emotions.

The visceral level is tightly coupled to the body's musculature—the motor system. This is what causes animals to fight or flee, or to relax. An animal's (or person's) visceral state can often be read by analyzing the tension of the body: tense means a negative state; relaxed, a positive state. Note, too, that we often determine our own body state by noting our own musculature. A common self-report might be something like, "I was tense, my fists clenched, and I was sweating."

Visceral responses are fast and completely subconscious. They are sensitive only to the current state of things. Most scientists do not call these emotions: they are precursors to emotion. Stand at the edge of a cliff and you will experience a visceral response. Or bask in the warm, comforting glow after a pleasant experience, perhaps a nice meal.

For designers, the visceral response is about immediate perception: the pleasantness of a mellow, harmonious sound or the jarring, irritating scratch of fingernails on a rough surface. Here is where the style matters: appearances, whether sound or sight, touch or smell, drive the visceral response. This has nothing to do with how usable, effective, or understandable the product is. It is all about attraction or repulsion. Great designers use their aesthetic sensibilities to drive these visceral responses.

Engineers and other logical people tend to dismiss the visceral response as irrelevant. Engineers are proud of the inherent quality of their work and dismayed when inferior products sell better "just because they look better." But all of us make these kinds of judgments, even those very logical engineers. That's why they love some of their tools and dislike others. Visceral responses matter.

THE BEHAVIORAL LEVEL

The *behavioral* level is the home of learned skills, triggered by situations that match the appropriate patterns. Actions and analyses at this level are largely subconscious. Even though we are usually aware of our actions, we are often unaware of the details. When we speak, we often do not know what we are about to say until our conscious mind (the reflective part of the mind) hears ourselves uttering the words. When we play a sport, we are prepared for action, but our responses occur far too quickly for conscious control: it is the behavioral level that takes control.

When we perform a well-learned action, all we have to do is think of the goal and the behavioral level handles all the details: the conscious mind has little or no awareness beyond creating the desire to act. It's actually interesting to keep trying it. Move the left hand, then the right. Stick out your tongue, or open your mouth. What did you do? You don't know. All you know is that you "willed" the action and the correct thing happened. You can even make the actions more complex. Pick up a cup, and then with the same hand, pick up several more items. You automatically adjust the fingers and the hand's orientation to make the task possible. You only need to pay conscious attention if the cup holds some liquid that you wish to avoid spilling. But even in that case, the actual control of the muscles is beneath conscious perception: concentrate on not spilling and the hands automatically adjust.

For designers, the most critical aspect of the behavioral level is that every action is associated with an expectation. Expect a positive outcome and the result is a positive affective response (a "positive valence," in the scientific literature). Expect a negative outcome and the result is a negative affective response (a negative valence): dread and hope, anxiety and anticipation. The information in the feedback loop of evaluation confirms or disconfirms the expectations, resulting in satisfaction or relief, disappointment or frustration.

Behavioral states are learned. They give rise to a feeling of control when there is good understanding and knowledge of results, and frustration and anger when things do not go as planned, and especially when neither the reason nor the possible remedies are known. Feedback provides reassurance, even when it indicates a negative result. A lack of feedback creates a feeling of lack of control, which can be unsettling. Feedback is critical to managing expectations, and good design provides this. Feedback—knowledge of results—is how expectations are resolved and is critical to learning and the development of skilled behavior.

Expectations play an important role in our emotional lives. This is why drivers tense when trying to get through an intersection before the light turns red, or students become highly anxious before an exam. The release of the tension of expectation creates a sense of relief. The emotional system is especially responsive to changes in states—so an upward change is interpreted positively even if it is only from a very bad state to a not-so-bad state, just as a change is interpreted negatively even if it is from an extremely positive state to one only somewhat less positive.

THE REFLECTIVE LEVEL

The *reflective* level is the home of conscious cognition. As a consequence, this is where deep understanding develops, where reasoning and conscious decision-making take place. The visceral and behavioral levels are subconscious and, as a result, they respond rapidly, but without much analysis. Reflection is cognitive, deep, and slow. It often occurs after the events have happened. It is a reflection or looking back over them, evaluating the circumstances, actions, and outcomes, often assessing blame or responsibility. The highest levels of emotions come from the reflective level, for it is here that causes are assigned and where predictions of the future take place. Adding causal elements to experienced events leads to such emotional states as guilt and pride (when we assume ourselves to be the cause) and blame and praise (when others are thought to be the cause). Most of us have probably experienced the extreme highs and lows of anticipated future events, all imagined by a runaway reflective cognitive system but intense enough to create the physiological responses associated with extreme anger or pleasure. Emotion and cognition are tightly intertwined.

DESIGN MUST TAKE PLACE AT ALL LEVELS: VISCERAL, BEHAVIORAL, AND REFLECTIVE

To the designer, reflection is perhaps the most important of the levels of processing. Reflection is conscious, and the emotions produced at this level are the most protracted: those that assign agency and cause, such as guilt and blame or praise and pride. Reflective responses are part of our memory of events. Memories last far longer than the immediate experience or the period of usage, which are the domains of the visceral and behavioral levels. It is reflection that drives us to recommend a product, to recommend that others use it—or perhaps to avoid it.

Reflective memories are often more important than reality. If we have a strongly positive visceral response but disappointing usability problems at the behavioral level, when we reflect back upon the product, the reflective level might very well weigh the positive response strongly enough to overlook the severe behavioral difficulties (hence the phrase, "Attractive things work better"). Similarly, too much frustration, especially toward the ending stage of use, and our reflections about the experience might overlook the positive visceral qualities. Advertisers hope that the strong reflective value associated with a well-known, highly prestigious brand might overwhelm our judgment, despite a frustrating experience in using the product. Vacations are often remembered with fondness, despite the evidence from diaries of repeated discomfort and anguish.

All three levels of processing work together. All play essential roles in determining a person's like or dislike of a product or service. One nasty experience with a service provider can spoil all future experiences. One superb experience can make up for past deficiencies. The behavioral level, which is the home of interaction, is also the home of all expectation-based emotions, of hope and joy, frustration and anger. Understanding arises at a combination of the behavioral and reflective levels. Enjoyment requires all three. Designing at all three levels is so important that I devote an entire book to the topic, *Emotional Design.*

In psychology, there has been a long debate about which happens first: emotion or cognition. Do we run and flee because some event happened that made us afraid? Or are we afraid because our conscious, reflective mind notices that we are running? The three-level analysis shows that both of these ideas can be correct. Sometimes the emotion comes first. An unexpected loud noise can cause automatic visceral and behavioral responses that make us flee. Then, the reflective system observes itself fleeing and deduces that it is afraid. The actions of running and fleeing occur first and set off the interpretation of fear.

But sometimes cognition occurs first. Suppose the street where we are walking leads to a dark and narrow section. Our reflective system might conjure numerous imagined threats that await us. At some point, the imagined depiction of potential harm is large enough to trigger the behavioral system, causing us to turn, run, and flee. Here is where the cognition sets off the fear and the action.

Most products do not cause fear, running, or fleeing, but badly designed devices can induce frustration and anger, a feeling of helplessness and despair, and possibly even hate. Well-designed devices can induce pride and enjoyment, a feeling of being in control and pleasure—possibly even love and attachment. Amusement parks are experts at balancing the conflicting responses of the emotional stages, providing rides and fun houses that trigger fear responses from the visceral and behavioral levels, while all the time providing reassurance at the reflective level that the park would never subject anyone to real danger.

All three levels of processing work together to determine a person's cognitive and emotional state. High-level reflective cognition can trigger lower-level emotions. Lower-level emotions can trigger higher-level reflective cognition.

The Seven Stages of Action and the Three Levels of Processing

The stages of action can readily be associated with the three different levels of processing, as shown in Figure 2.4. At the lowest level are the visceral levels of calmness or anxiety when approaching a task or evaluating the state of the world. Then, in the middle level, are the behavioral ones driven by expectations on the execution side—for example, hope and fear—and emotions driven by the confirmation of those expectations on the evaluation side—for example, relief or despair. At the highest level are the reflective emotions, ones that assess the results in terms of the presumed causal agents and the consequences, both immediate and long-term. Here is where satisfaction and pride occur, or perhaps blame and anger.

One important emotional state is the one that accompanies complete immersion into an activity, a state that the social scientist Mihaly Csikszentmihalyi has labeled "flow." Csikszentmihalyi has long studied how people interact with their work and play, and how their lives reflect this intermix of activities. When in the flow state, people lose track of time and the outside environment.

They are at one with the task they are performing. The task, moreover, is at just the proper level of difficulty: difficult enough to provide a challenge and require continued attention, but not so difficult that it invokes frustration and anxiety.

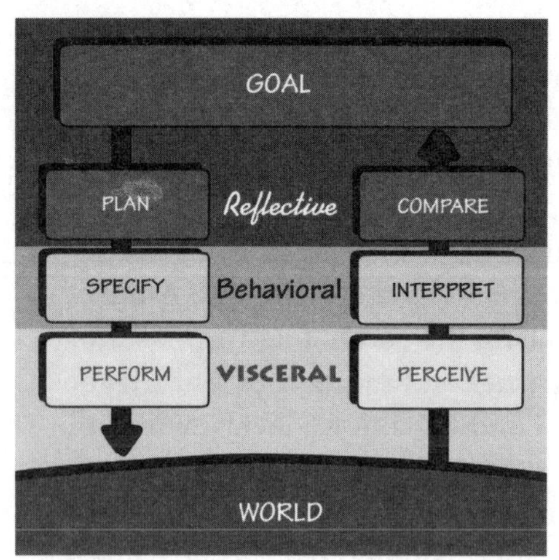

FIGURE 2.4. Levels of Processing and the Stages of the Action Cycle. Visceral response is at the lowest level: the control of simple muscles and sensing the state of the world and body. The behavioral level is about expectations, so it is sensitive to the expectations of the action sequence and then the interpretations of the feedback. The reflective level is a part of the goal- and plan-setting activity as well as affected by the comparison of expectations with what has actually happened.

Csikszentmihalyi's work shows how the behavioral level creates a powerful set of emotional responses. Here, the subconscious expectations established by the execution side of the action cycle set up emotional states dependent upon those expectations. When the results of our actions are evaluated against expectations, the resulting emotions affect our feelings as we continue through the many cycles of action. An easy task, far below our skill level, makes it so easy to meet expectations that there is no challenge. Very little or no processing effort is required, which leads to apathy or boredom. A difficult task, far above our skill, leads to so many failed expectations that it causes frustration, anxiety, and helplessness. The flow state occurs when the challenge of the activity just slightly exceeds our skill level, so full attention is continually required. Flow requires that the activity be neither too easy nor too difficult relative to our level of skill. The constant tension coupled with continual progress and success can be an engaging, immersive experience sometimes lasting for hours.

People as Storytellers

Now that we have explored the way that actions get done and the three different levels of processing that integrate cognition and emotion, we are ready to look at some of the implications.

People are innately disposed to look for causes of events, to form explanations and stories. That is one reason storytelling is such a persuasive medium. Stories resonate with our experiences and provide examples of new instances. From our experiences and the stories of others we tend to form generalizations about the way people behave and things work. We attribute causes to events, and as long as these cause-and-effect pairings make sense, we accept them and use them for understanding future events. Yet these causal attributions are often erroneous. Sometimes they implicate the wrong causes, and for some things that happen, there is no single cause; rather, a complex chain of events that all contribute to the result: if any one of the events would not have occurred, the result would be different. But even when there is no single causal act, that doesn't stop people from assigning one.

Conceptual models are a form of story, resulting from our predisposition to find explanations. These models are essential in helping us understand our experiences, predict the outcome of our actions, and handle unexpected occurrences. We base our models on whatever knowledge we have, real or imaginary, naive or sophisticated.

Conceptual models are often constructed from fragmentary evidence, with only a poor understanding of what is happening, and with a kind of naive psychology that postulates causes, mechanisms, and relationships even where there are none. Some faulty models lead to the frustrations of everyday life, as in the case of my unsettable refrigerator, where my conceptual model of its operation (see again Figure 1.10A) did not correspond to reality (Figure 1.10B). Far more serious are faulty models of such complex systems as an industrial plant or passenger airplane. Misunderstanding there can lead to devastating accidents.

Consider the thermostat that controls room heating and cooling systems. How does it work? The average thermostat offers almost no evidence of its operation except in a highly roundabout manner. All we know is that if the room is too cold, we set a higher temperature into the thermostat. Eventually we feel warmer. Note that the same thing applies to the temperature control for almost any device whose temperature is to be regulated. Want to bake a cake? Set the oven thermostat and the oven goes to the desired temperature.

If you are in a cold room, in a hurry to get warm, will the room heat more quickly if you turn the thermostat to its maximum setting? Or if you want the oven to reach its working temperature faster, should you turn the temperature dial all the way to maximum, then turn it down once the desired temperature is reached? Or to cool a room most quickly, should you set the air conditioner thermostat to its lowest temperature setting?

If you think that the room or oven will cool or heat faster if the thermostat is turned all the way to the maximum setting, you are wrong—you hold an erroneous folk theory of the heating and cooling system. One commonly held folk theory of the working of a thermostat is that it is like a valve: the thermostat controls how much heat (or cold) comes out of the device. Hence, to heat or cool something most quickly, set the thermostat so that the device is on maximum. The theory is reasonable, and there exist devices that operate like this, but neither the heating or cooling equipment for a home nor the heating element of a traditional oven is one of them.

In most homes, the thermostat is just an on-off switch. Moreover, most heating and cooling devices are either fully on or fully off: all or nothing, with no in-between states. As a result, the thermostat turns the heater, oven, or air conditioner completely on, at full power, until the temperature setting on the thermostat is reached. Then it turns the unit completely off. Setting the thermostat at one extreme cannot affect how long it takes to reach the desired temperature. Worse, because this bypasses the automatic shutoff when the desired temperature is reached, setting it at the extremes invariably means that the temperature overshoots the target. If people were uncomfortably cold or hot before, they will become uncomfortable in the other direction, wasting considerable energy in the process.

But how are you to know? What information helps you understand how the thermostat works? The design problem with the refrigerator is that there are no aids to understanding, no way of forming the correct conceptual model. In fact, the information provided misleads people into forming the wrong, quite inappropriate model.

The real point of these examples is not that some people have erroneous beliefs; it is that everyone forms stories (conceptual models) to explain what they have observed. In the absence of external information, people can let their imagination run free as long as the conceptual models they develop account for the facts as they perceive them. As a result, people use their thermostats inappropriately, causing themselves unnecessary effort, and often resulting in large temperature swings, thus wasting energy, which is both a needless expense and bad for the environment. (Later in this chapter, page 69, I provide an example of a thermostat that does provide a useful conceptual model.)

Blaming the Wrong Things

People try to find causes for events. They tend to assign a causal relation whenever two things occur in succession. If some unexpected event happens in my home just after I have taken some action, I am apt to conclude that it was caused by that action, even if there really was no relationship between the two. Similarly, if I do something expecting a result and nothing happens, I am apt to interpret this lack of informative feedback as an indication that I didn't do the action correctly: the most likely thing to do, therefore, is to repeat the action, only with more force. Push a door and it fails to open? Push again, harder. With electronic devices, if the feedback is delayed sufficiently, people often are led to conclude that the press wasn't recorded, so they do the same action again, sometimes repeatedly, unaware that all of their presses were recorded. This can lead to unintended results. Repeated presses might intensify the response much more than was intended. Alternatively, a second request might cancel the previous one, so that an odd number of pushes produces the desired result, whereas an even number leads to no result.

The tendency to repeat an action when the first attempt fails can be disastrous. This has led to numerous deaths when people tried to escape a burning building by attempting to push open exit doors that opened inward, doors that should have been pulled. As a result, in many countries, the law requires doors in public places to open outward, and moreover to be operated by so-called panic bars, so that they automatically open when people, in a panic to escape a fire, push their bodies against them. This is a great application of appropriate affordances: see the door in Figure 2.5.

Modern systems try hard to provide feedback within 0.1 second of any operation, to reassure the user that the request was received. This is especially important if the operation will take considerable time. The presence of a filling hourglass or rotating clock hands is a reassuring sign that work is in progress. When the delay can be predicted, some systems provide time estimates as well as progress bars to indicate how far along the task has gone. More systems should adopt these sensible displays to provide timely and meaningful feedback of results.

Some studies show it is wise to underpredict—that is, to say an operation will take longer than it actually will. When the system computes the amount of time, it can compute the range of possible

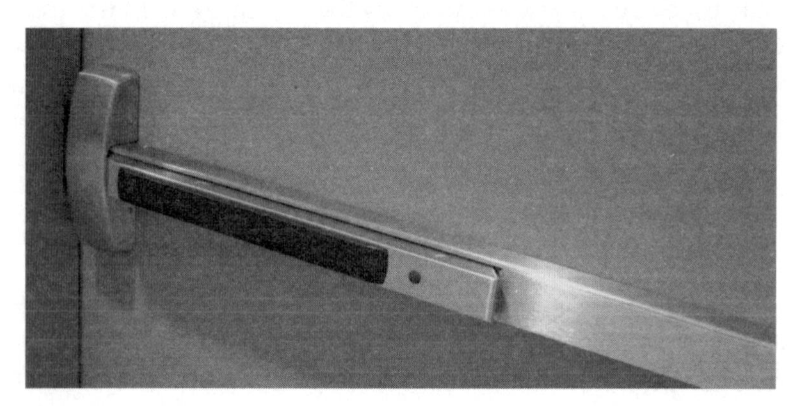

FIGURE 2.5. **Panic Bars on Doors.** People fleeing a fire would die if they encountered exit doors that opened inward, because they would keep trying to push them outward, and when that failed, they would push harder. The proper design, now required by law in many places, is to change the design of doors so that they open when pushed. Here is one example: an excellent design strategy for dealing with real behavior by the use of the proper affordances coupled with a graceful signifier, the black bar, which indicates where to push. (Photograph by author at the Ford Design Center, Northwestern University.)

times. In that case it ought to display the range, or if only a single value is desirable, show the slowest, longest value. That way, the expectations are liable to be exceeded, leading to a happy result.

When it is difficult to determine the cause of a difficulty, where do people put the blame? Often people will use their own conceptual models of the world to determine the perceived causal relationship between the thing being blamed and the result. The word *perceived* is critical: the causal relationship does not have to exist; the person simply has to think it is there. Sometimes the result is to attribute cause to things that had nothing to do with the action.

Suppose I try to use an everyday thing, but I can't. Who is at fault: me or the thing? We are apt to blame ourselves, especially if others are able to use it. Suppose the fault really lies in the device, so that lots of people have the same problems. Because everyone perceives the fault to be his or her own, nobody wants to admit to having trouble. This creates a conspiracy of silence, where the feelings of guilt and helplessness among people are kept hidden.

Interestingly enough, the common tendency to blame ourselves for failures with everyday objects goes against the normal attributions we make about ourselves and others. Everyone sometimes acts in a way that seems strange, bizarre, or simply wrong and inappropriate. When we do this, we tend to attribute our behavior to the environment. When we see others do it, we tend to attribute it to their personalities.

Here is a made-up example. Consider Tom, the office terror. Today, Tom got to work late, yelled at his colleagues because the office coffee machine was empty, then ran to his office and slammed the door shut. "Ah," his colleagues and staff say to one another, "there he goes again."

Now consider Tom's point of view. "I really had a hard day," Tom explains. "I woke up late because my alarm clock failed to go off: I didn't even have time for my morning coffee. Then I couldn't find a parking spot because I was late. And there wasn't any coffee in the office machine; it was all out. None of this was my fault—I had a run of really bad events. Yes, I was a bit curt, but who wouldn't be under the same circumstances?"

Tom's colleagues don't have access to his inner thoughts or to his morning's activities. All they see is that Tom yelled at them simply because the office coffee machine was empty. This reminds them of another similar event. "He does that all the time," they conclude, "always blowing up over the most minor things." Who is correct? Tom or his colleagues? The events can be seen from two different points of view with two different interpretations: common responses to the trials of life or the result of an explosive, irascible personality.

It seems natural for people to blame their own misfortunes on the environment. It seems equally natural to blame other people's misfortunes on their personalities. Just the opposite attribution, by the way, is made when things go well. When things go right, people credit their own abilities and intelligence. The onlookers do the reverse. When they see things go well for someone else, they sometimes credit the environment, or luck.

In all such cases, whether a person is inappropriately accepting blame for the inability to work simple objects or attributing behavior to environment or personality, a faulty conceptual model is at work.

LEARNED HELPLESSNESS

The phenomenon called *learned helplessness* might help explain the self-blame. It refers to the situation in which people experience repeated failure at a task. As a result, they decide that the task cannot be done, at least not by them: they are helpless. They stop trying. If this feeling covers a group of tasks, the result can be severe difficulties coping with life. In the extreme case, such learned helplessness leads to depression and to a belief that the individuals cannot cope with everyday life at all. Sometimes all it takes to get such a feeling of helplessness are a few experiences that accidentally turn out bad. The phenomenon has been most frequently studied as a precursor to the clinical problem of depression, but I have seen it happen after a few bad experiences with everyday objects.

Do common technology and mathematics phobias result from a kind of learned helplessness? Could a few instances of failure in what appear to be straightforward situations generalize to every technological object, every mathematics problem? Perhaps. In fact, the design of everyday things (and the design of mathematics courses) seems almost guaranteed to cause this. We could call this phenomenon taught helplessness.

When people have trouble using technology, especially when they perceive (usually incorrectly) that nobody else is having the same problems, they tend to blame themselves. Worse, the more they have trouble, the more helpless they may feel, believing that they must be technically or mechanically inept. This is just the opposite of the more normal situation where people blame their own difficulties on the environment. This false blame is especially ironic because the culprit here is usually the poor design of the technology, so blaming the environment (the technology) would be completely appropriate.

Consider the normal mathematics curriculum, which continues relentlessly on its way, each new lesson assuming full knowledge and understanding of all that has passed before. Even though each point may be simple, once you fall behind it is hard to catch up. The result: mathematics phobia—not because the material is difficult, but because it is taught so that difficulty in one stage hinders further progress. The problem is that once failure starts, it is soon generalized by self-blame to all of mathematics. Similar processes are at work with technology. The vicious cycle starts: if you fail at something, you think it is your fault. Therefore you think you can't do that task. As a result, next time you have to do the task, you believe you can't, so you don't even try. The result is that you can't, just as you thought.

You're trapped in a self-fulfilling prophecy.

POSITIVE PSYCHOLOGY

Just as we learn to give up after repeated failure, we can learn optimistic, positive responses to life. For years, psychologists focused upon the gloomy story of how people failed, on the limits of human abilities, and on psychopathologies—depression, mania, paranoia, and so on. But the twenty-first century sees a new approach:

to focus upon a positive psychology, a culture of positive thinking, of feeling good about oneself. In fact, the normal emotional state of most people is positive. When something doesn't work, it can be considered an interesting challenge, or perhaps just a positive learning experience.

We need to remove the word *failure* from our vocabulary, replacing it instead with *learning experience.* To fail is to learn: we learn more from our failures than from our successes. With success, sure, we are pleased, but we often have no idea why we succeeded. With failure, it is often possible to figure out why, to ensure that it will never happen again.

Scientists know this. Scientists do experiments to learn how the world works. Sometimes their experiments work as expected, but often they don't. Are these failures? No, they are learning experiences. Many of the most important scientific discoveries have come from these so-called failures.

Failure can be such a powerful learning tool that many designers take pride in their failures that happen while a product is still in development. One design firm, IDEO, has it as a creed: "Fail often, fail fast," they say, for they know that each failure teaches them a lot about what to do right. Designers need to fail, as do researchers. I have long held the belief—and encouraged it in my students and employees—that failures are an essential part of exploration and creativity. If designers and researchers do not sometimes fail, it is a sign that they are not trying hard enough—they are not thinking the great creative thoughts that will provide breakthroughs in how we do things. It is possible to avoid failure, to always be safe. But that is also the route to a dull, uninteresting life.

The designs of our products and services must also follow this philosophy. So, to the designers who are reading this, let me give some advice:

- Do not blame people when they fail to use your products properly.
- Take people's difficulties as signifiers of where the product can be improved.

- Eliminate all error messages from electronic or computer systems. Instead, provide help and guidance.
- Make it possible to correct problems directly from help and guidance messages. Allow people to continue with their task: Don't impede progress—help make it smooth and continuous. Never make people start over.
- Assume that what people have done is partially correct, so if it is inappropriate, provide the guidance that allows them to correct the problem and be on their way.
- Think positively, for yourself and for the people you interact with.

Falsely Blaming Yourself

I have studied people making errors—sometimes serious ones—with mechanical devices, light switches and fuses, computer operating systems and word processors, even airplanes and nuclear power plants. Invariably people feel guilty and either try to hide the error or blame themselves for "stupidity" or "clumsiness." I often have difficulty getting permission to watch: nobody likes to be observed performing badly. I point out that the design is faulty and that others make the same errors, yet if the task appears simple or trivial, people still blame themselves. It is almost as if they take perverse pride in thinking of themselves as mechanically incompetent.

I once was asked by a large computer company to evaluate a brand-new product. I spent a day learning to use it and trying it out on various problems. In using the keyboard to enter data, it was necessary to differentiate between the Return key and the Enter key. If the wrong key was pressed, the last few minutes' work was irrevocably lost.

I pointed out this problem to the designer, explaining that I, myself, had made the error frequently and that my analyses indicated that this was very likely to be a frequent error among users. The designer's first response was: "Why did you make that error? Didn't you read the manual?" He proceeded to explain the different functions of the two keys.

"Yes, yes," I explained, "I understand the two keys, I simply confuse them. They have similar functions, are located in similar locations on the keyboard, and as a skilled typist, I often hit Return automatically, without thought. Certainly others have had similar problems."

"Nope," said the designer. He claimed that I was the only person who had ever complained, and the company's employees had been using the system for many months. I was skeptical, so we went together to some of the employees and asked them whether they had ever hit the Return key when they should have hit Enter. And did they ever lose their work as a result?

"Oh, yes," they said, "we do that a lot."

Well, how come nobody ever said anything about it? After all, they were encouraged to report all problems with the system. The reason was simple: when the system stopped working or did something strange, they dutifully reported it as a problem. But when they made the Return versus Enter error, they blamed themselves. After all, they had been told what to do. They had simply erred.

The idea that a person is at fault when something goes wrong is deeply entrenched in society. That's why we blame others and even ourselves. Unfortunately, the idea that a person is at fault is imbedded in the legal system. When major accidents occur, official courts of inquiry are set up to assess the blame. More and more often the blame is attributed to "human error." The person involved can be fined, punished, or fired. Maybe training procedures are revised. The law rests comfortably. But in my experience, human error usually is a result of poor design: it should be called system error. Humans err continually; it is an intrinsic part of our nature. System design should take this into account. Pinning the blame on the person may be a comfortable way to proceed, but why was the system ever designed so that a single act by a single person could cause calamity? Worse, blaming the person without fixing the root, underlying cause does not fix the problem: the same error is likely to be repeated by someone else. I return to the topic of human error in Chapter 5.

Of course, people do make errors. Complex devices will always require some instruction, and someone using them without instruction should expect to make errors and to be confused. But designers should take special pains to make errors as cost-free as possible. Here is my credo about errors:

> Eliminate the term *human error*. Instead, talk about communication and interaction: what we call an error is usually bad communication or interaction. When people collaborate with one another, the word error is never used to characterize another person's utterance. That's because each person is trying to understand and respond to the other, and when something is not understood or seems inappropriate, it is questioned, clarified, and the collaboration continues. Why can't the interaction between a person and a machine be thought of as collaboration?
>
> Machines are not people. They can't communicate and understand the same way we do. This means that their designers have a special obligation to ensure that the behavior of machines is understandable to the people who interact with them. True collaboration requires each party to make some effort to accommodate and understand the other. When we collaborate with machines, it is people who must do all the accommodation. Why shouldn't the machine be more friendly? The machine should accept normal human behavior, but just as people often subconsciously assess the accuracy of things being said, machines should judge the quality of information given it, in this case to help its operators avoid grievous errors because of simple slips (discussed in Chapter 5). Today, we insist that people perform abnormally, to adapt themselves to the peculiar demands of machines, which includes always giving precise, accurate information. Humans are particularly bad at this, yet when they fail to meet the arbitrary, inhuman requirements of machines, we call it human error. No, it is design error.
>
> Designers should strive to minimize the chance of inappropriate actions in the first place by using affordances, signifiers, good mapping, and constraints to guide the actions. If a person performs an inappropriate action, the design should maximize the chance that this can be discovered and then rectified. This requires good, intelligible feedback coupled with a simple, clear conceptual model. When people understand what has happened, what state the system is in, and what the most appropriate set of actions is, they can perform their activities more effectively.

People are not machines. Machines don't have to deal with continual interruptions. People are subjected to continual interruptions. As a result, we are often bouncing back and forth between tasks, having to recover our place, what we were doing, and what we were thinking when we return to a previous task. No wonder we sometimes forget our place when we return to the original task, either skipping or repeating a step, or imprecisely retaining the information we were about to enter.

Our strengths are in our flexibility and creativity, in coming up with solutions to novel problems. We are creative and imaginative, not mechanical and precise. Machines require precision and accuracy; people don't. And we are particularly bad at providing precise and accurate inputs. So why are we always required to do so? Why do we put the requirements of machines above those of people?

When people interact with machines, things will not always go smoothly. This is to be expected. So designers should anticipate this. It is easy to design devices that work well when everything goes as planned. The hard and necessary part of design is to make things work well even when things do not go as planned.

HOW TECHNOLOGY CAN ACCOMMODATE HUMAN BEHAVIOR

In the past, cost prevented many manufacturers from providing useful feedback that would assist people in forming accurate conceptual models. The cost of color displays large and flexible enough to provide the required information was prohibitive for small, inexpensive devices. But as the cost of sensors and displays has dropped, it is now possible to do a lot more.

Thanks to display screens, telephones are much easier to use than ever before, so my extensive criticisms of phones found in the earlier edition of this book have been removed. I look forward to great improvements in all our devices now that the importance of these design principles are becoming recognized and the enhanced quality and lower costs of displays make it possible to implement the ideas.

PROVIDING A CONCEPTUAL MODEL FOR A HOME THERMOSTAT

My thermostat, for example (designed by Nest Labs), has a colorful display that is normally off, turning on only when it senses that I

FIGURE 2.6. A Thermostat with an Explicit Conceptual Model. This thermostat, manufactured by Nest Labs, helps people form a good conceptual model of its operation. Photo A shows the thermostat. The background, blue, indicates that it is now cooling the home. The current temperature is 75°F (24°C) and the target temperature is 72°F (22°C), which it expects to reach in 20 minutes. Photo B shows its use of a smart phone to deliver a summary of its settings and the home's energy use. Both A and B combine to help the home dweller develop conceptual models of the thermostat and the home's energy consumption. (Photographs courtesy of Nest Labs, Inc.)

A.

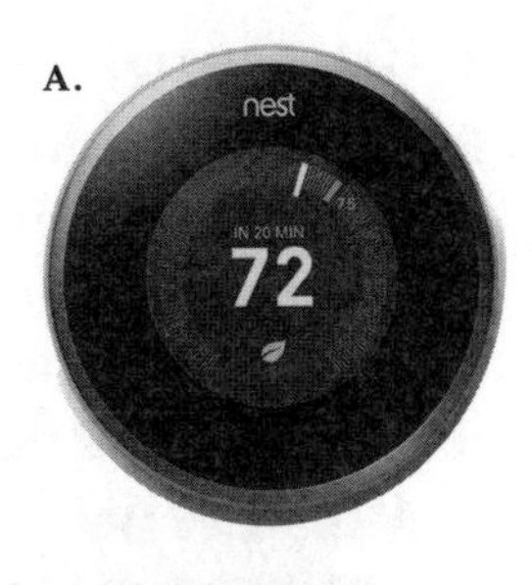

B.

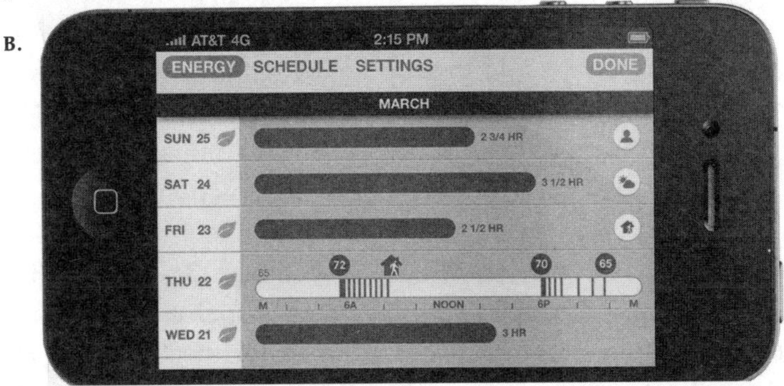

am nearby. Then it provides me with the current temperature of the room, the temperature to which it is set, and whether it is heating or cooling the room (the background color changes from black when it is neither heating nor cooling, to orange while heating, or to blue while cooling). It learns my daily patterns, so it changes temperature automatically, lowering it at bedtime, raising it again in the morning, and going into "away" mode when it detects that nobody is in the house. All the time, it explains what it is doing. Thus, when it has to change the room temperature substantially (either because someone has entered a manual change or because it has decided that it is now time to switch), it gives a prediction: "Now 75°, will be 72° in 20 minutes." In addition, Nest can be connected wirelessly to smart devices that allow for remote operation of the thermostat and also for larger screens to provide a detailed analysis of its performance, aiding the home occupant's development of a conceptual model both of Nest and also of the home's energy consumption. Is Nest perfect? No, but it marks improvement in the collaborative interaction of people and everyday things.

ENTERING DATES, TIMES, AND TELEPHONE NUMBERS

Many machines are programmed to be very fussy about the form of input they require, where the fussiness is not a requirement of the machine but due to the lack of consideration for people in the design of the software. In other words: inappropriate programming. Consider these examples.

Many of us spend hours filling out forms on computers—forms that require names, dates, addresses, telephone numbers, monetary sums, and other information in a fixed, rigid format. Worse, often we are not even told the correct format until we get it wrong. Why not figure out the variety of ways a person might fill out a form and accommodate all of them? Some companies have done excellent jobs at this, so let us celebrate their actions.

Consider Microsoft's calendar program. Here, it is possible to specify dates any way you like: "November 23, 2015," "23 Nov. 15," or "11.23.15." It even accepts phrases such as "a week from Thursday," "tomorrow," "a week from tomorrow," or "yesterday." Same with time. You can enter the time any way you want: "3:45 PM," "15.35," "an hour," "two and one-half hours." Same with telephone numbers: Want to start with a + sign (to indicate the code for international dialing)? No problem. Like to separate the number fields with spaces, dashes, parentheses, slashes, periods? No problem. As long as the program can decipher the date, time, or telephone number into a legal format, it is accepted. I hope the team that worked on this got bonuses and promotions.

Although I single out Microsoft for being the pioneer in accepting a wide variety of formats, it is now becoming standard practice. By the time you read this, I would hope that every program would permit any intelligible format for names, dates, phone numbers, street addresses, and so on, transforming whatever is entered into whatever form the internal programming needs. But I predict that even in the twenty-second century, there will still be forms that require precise accurate (but arbitrary) formats for no reason except the laziness of the programming team. Perhaps in the years that pass between this book's publication and when you are reading this, great improvements will have been made. If we are all lucky, this section will be badly out of date. I hope so.

The Seven Stages of Action: Seven Fundamental Design Principles

The seven-stage model of the action cycle can be a valuable design tool, for it provides a basic checklist of questions to ask. In general, each stage of action requires its own special design strategies and, in turn, provides its own opportunity for disaster. Figure 2.7 summarizes the questions:

1. What do I want to accomplish?
2. What are the alternative action sequences?
3. What action can I do now?
4. How do I do it?
5. What happened?
6. What does it mean?
7. Is this okay? Have I accomplished my goal?

Anyone using a product should always be able to determine the answers to all seven questions. This puts the burden on the designer

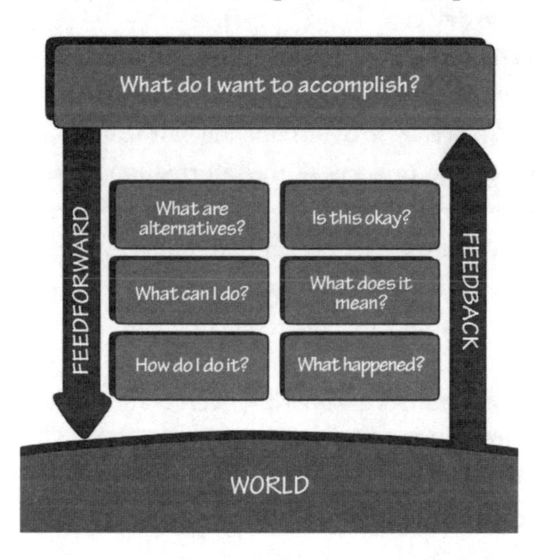

FIGURE 2.7. The Seven Stages of Action as Design Aids. Each of the seven stages indicates a place where the person using the system has a question. The seven questions pose seven design themes. How should the design convey the information required to answer the user's question? Through appropriate constraint and mappings, signifiers and conceptual models, feedback and visibility. The information that helps answer questions of execution (doing) is *feedforward*. The information that aids in understanding what has happened is *feedback*.

to ensure that at each stage, the product provides the information required to answer the question.

The information that helps answer questions of execution (doing) is *feedforward*. The information that aids in understanding what has happened is *feedback*. Everyone knows what feedback is. It helps you know what happened. But how do you know what you can do? That's the role of feedforward, a term borrowed from control theory.

Feedforward is accomplished through appropriate use of signifiers, constraints, and mappings. The conceptual model plays an important role. Feedback is accomplished through explicit information about the impact of the action. Once again, the conceptual model plays an important role.

Both feedback and feedforward need to be presented in a form that is readily interpreted by the people using the system. The presentation has to match how people view the goal they are trying to achieve and their expectations. Information must match human needs.

The insights from the seven stages of action lead us to seven fundamental principles of design:

1. **Discoverability.** It is possible to determine what actions are possible and the current state of the device.
2. **Feedback.** There is full and continuous information about the results of actions and the current state of the product or service. After an action has been executed, it is easy to determine the new state.
3. **Conceptual model.** The design projects all the information needed to create a good conceptual model of the system, leading to understanding and a feeling of control. The conceptual model enhances both discoverability and evaluation of results.
4. **Affordances.** The proper affordances exist to make the desired actions possible.
5. **Signifiers.** Effective use of signifiers ensures discoverability and that the feedback is well communicated and intelligible.
6. **Mappings.** The relationship between controls and their actions follows the principles of good mapping, enhanced as much as possible through spatial layout and temporal contiguity.

7. **Constraints.** Providing physical, logical, semantic, and cultural constraints guides actions and eases interpretation.

The next time you can't immediately figure out the shower control in a hotel room or have trouble using an unfamiliar television set or kitchen appliance, remember that the problem is in the design. Ask yourself where the problem lies. At which of the seven stages of action does it fail? Which design principles are deficient?

But it is easy to find fault: the key is to be able to do things better. Ask yourself how the difficulty came about. Realize that many different groups of people might have been involved, each of which might have had intelligent, sensible reasons for their actions. For example, a troublesome bathroom shower was designed by people who were unable to know how it would be installed, then the shower controls might have been selected by a building contractor to fit the home plans provided by yet another person. Finally, a plumber, who may not have had contact with any of the other people, did the installation. Where did the problems arise? It could have been at any one (or several) of these stages. The result may appear to be poor design, but it may actually arise from poor communication.

One of my self-imposed rules is, "Don't criticize unless you can do better." Try to understand how the faulty design might have occurred: try to determine how it could have been done otherwise. Thinking about the causes and possible fixes to bad design should make you better appreciate good design. So, the next time you come across a well-designed object, one that you can use smoothly and effortlessly on the first try, stop and examine it. Consider how well it masters the seven stages of action and the principles of design. Recognize that most of our interactions with products are actually interactions with a complex system: good design requires consideration of the entire system to ensure that the requirements, intentions, and desires at each stage are faithfully understood and respected at all the other stages.

CHAPTER THREE

KNOWLEDGE IN THE HEAD AND IN THE WORLD

A friend kindly let me borrow his car, an older, classic Saab. Just before I was about to leave, I found a note waiting for me: "I should have mentioned that to get the key out of the ignition, the car needs to be in reverse." The car needs to be in reverse! If I hadn't seen the note, I never could have figured that out. There was no visible cue in the car: the knowledge needed for this trick had to reside in the head. If the driver lacks that knowledge, the key stays in the ignition forever.

Every day we are confronted by numerous objects, devices, and services, each of which requires us to behave or act in some particular manner. Overall, we manage quite well. Our knowledge is often quite incomplete, ambiguous, or even wrong, but that doesn't matter: we still get through the day just fine. How do we manage? We combine knowledge in the head with knowledge in the world. Why combine? Because neither alone will suffice.

It is easy to demonstrate the faulty nature of human knowledge and memory. The psychologists Ray Nickerson and Marilyn Adams showed that people do not remember what common coins look like (Figure 3.1). Even though the example is for the American one-cent piece, the penny, the finding holds true for currencies across the world. But despite our ignorance of the coins' appearance, we use our money properly.

Why the apparent discrepancy between the precision of behavior and the imprecision of knowledge? Because not all of the knowl-

FIGURE 3.1. Which Is the US One-Cent Coin, the Penny? Fewer than half of the American college students who were given this set of drawings and asked to select the correct image could do so. Pretty bad performance, except that the students, of course, have no difficulty using the money. In normal life, we have to distinguish between the penny and other coins, not among several versions of one denomination. Although this is an old study using American coins, the results still hold true today using coins of any currency. (From Nickerson & Adams, 1979, *Cognitive Psychology*, 11 (3). Reproduced with permission of Academic Press via Copyright Clearance Center.)

edge required for precise behavior has to be in the head. It can be distributed—partly in the head, partly in the world, and partly in the constraints of the world.

Precise Behavior from Imprecise Knowledge

Precise behavior can emerge from imprecise knowledge for four reasons:

1. **Knowledge is both in the head and in the world.** Technically, knowledge can only be in the head, because knowledge requires interpretation and understanding, but once the world's structure has been interpreted and understood, it counts as knowledge. Much of the knowledge a person needs to do a task can be derived from the information in the world. Behavior is determined by combining the knowledge in the head with that in the world. For this chapter, I will use the term "knowledge" for both what is in the head and what is in the world. Although technically imprecise, it simplifies the discussion and understanding.

2. **Great precision is not required.** Precision, accuracy, and completeness of knowledge are seldom required. Perfect behavior results if the combined knowledge in the head and in the world is sufficient to distinguish an appropriate choice from all others.
3. **Natural constraints exist in the world.** The world has many natural, physical constraints that restrict the possible behavior: such things as the order in which parts can go together and the ways by which an object can be moved, picked up, or otherwise manipulated. This is knowledge in the world. Each object has physical features—projections, depressions, screw threads, appendages—that limit its relationships with other objects, the operations that can be performed on it, what can be attached to it, and so on.
4. **Knowledge of cultural constraints and conventions exists in the head.** Cultural constraints and conventions are learned artificial restrictions on behavior that reduce the set of likely actions, in many cases leaving only one or two possibilities. This is knowledge in the head. Once learned, these constraints apply to a wide variety of circumstances.

Because behavior can be guided by the combination of internal and external knowledge and constraints, people can minimize the amount of material they must learn, as well as the completeness, precision, accuracy, or depth of the learning. They also can deliberately organize the environment to support behavior. This is how nonreaders can hide their inability, even in situations where their job requires reading skills. People with hearing deficits (or with normal hearing but in noisy environments) learn to use other cues. Many of us manage quite well when in novel, confusing situations where we do not know what is expected of us. How do we do this? We arrange things so that we do not need to have complete knowledge or we rely upon the knowledge of the people around us, copying their behavior or getting them to do the required actions. It is actually quite amazing how often it is possible to hide one's ignorance, to get by without understanding or even much interest.

Although it is best when people have considerable knowledge and experience using a particular product—knowledge in the head—

the designer can put sufficient cues into the design—knowledge in the world—that good performance results even in the absence of previous knowledge. Combine the two, knowledge in the head and in the world, and performance is even better. How can the designer put knowledge into the device itself?

Chapters 1 and 2 introduced a wide range of fundamental design principles derived from research on human cognition and emotion. This chapter shows how knowledge in the world combines with knowledge in the head. Knowledge in the head is knowledge in the human memory system, so this chapter contains a brief review of the critical aspects of memory necessary for the design of usable products. I emphasize that for practical purposes, we do not need to know the details of scientific theories but simpler, more general, useful approximations. Simplified models are the key to successful application. The chapter concludes with a discussion of how natural mappings present information in the world in a manner readily interpreted and usable.

KNOWLEDGE IS IN THE WORLD

Whenever knowledge needed to do a task is readily available in the world, the need for us to learn it diminishes. For example, we lack knowledge about common coins, even though we recognize them just fine (Figure 3.1). In knowing what our currency looks like, we don't need to know all the details, simply sufficient knowledge to distinguish one value of currency from another. Only a small minority of people must know enough to distinguish counterfeit from legitimate money.

Or consider typing. Many typists have not memorized the keyboard. Usually each key is labeled, so nontypists can hunt and peck letter by letter, relying on knowledge in the world and minimizing the time required for learning. The problem is that such typing is slow and difficult. With experience, of course, hunt-and-peckers learn the positions of many of the letters on the keyboard, even without instruction, and typing speed increases notably, quickly surpassing handwriting speeds and, for some, reaching quite respectable rates. Peripheral vision and the feel of the keyboard provide some knowledge about key locations. Frequently used keys become completely learned, infrequently used keys are not learned well, and the other keys are partially learned. But as long as a typist needs to watch the keyboard, the speed is limited. The knowledge is still mostly in the world, not in the head.

If a person needs to type large amounts of material regularly, further investment is worthwhile: a course, a book, or an interactive program. The important thing is to learn the proper placement of fingers on the keyboard, to learn to type without looking, to get knowledge about the keyboard from the world into the head. It takes a few weeks to learn the system and several months of practice to become expert. But the payoff for all this effort is increased typing speed, increased accuracy, and decreased mental load and effort at the time of typing.

We only need to remember sufficient knowledge to let us get our tasks done. Because so much knowledge is available in the environment, it is surprising how little we need to learn. This is one reason people can function well in their environment and still be unable to describe what they do.

People function through their use of two kinds of knowledge: knowledge *of* and knowledge *how*. Knowledge *of*—what psychologists call *declarative knowledge*—includes the knowledge of facts and rules. "Stop at red traffic lights." "New York City is north of Rome." "China has twice as many people as India." "To get the key out of the ignition of a Saab car, the gearshift must be in reverse." Declarative knowledge is easy to write and to teach. Note that knowledge of the rules does not mean they are followed. The drivers in many cities are often quite knowledgeable about the official driving regulations, but they do not necessarily obey them. Moreover, the knowledge does not have to be true. New York City is actually south of Rome. China has only slightly more people than India (roughly 10 percent). People may know many things: that doesn't mean they are true.

Knowledge *how*—what psychologists call *procedural knowledge*—is the knowledge that enables a person to be a skilled musician, to return a serve in tennis, or to move the tongue properly when saying the phrase "frightening witches." Procedural knowledge is difficult or impossible to write down and difficult to teach. It is best taught by demonstration and best learned through practice. Even the best teachers cannot usually describe what they are doing. Procedural knowledge is largely subconscious, residing at the behavioral level of processing.

Knowledge in the world is usually easy to come by. Signifiers, physical constraints, and natural mappings are all perceivable cues that act as knowledge in the world. This type of knowledge occurs so commonly that we take it for granted. It is everywhere: the locations of letters on a keyboard; the lights and labels on controls that remind us of their purpose and give information about the current state of the device. Industrial equipment is replete with signal lights, indicators, and other reminders. We make extensive use of written notes. We place items in specific locations as reminders. In general, people structure their environment to provide a considerable amount of the knowledge required for something to be remembered.

Many organize their lives spatially in the world, creating a pile here, a pile there, each indicating some activity to be done, some event in progress. Probably everybody uses such a strategy to some extent. Look around you at the variety of ways people arrange their rooms and desks. Many styles of organization are possible, but invariably the physical layout and visibility of the items convey information about relative importance.

WHEN PRECISION IS UNEXPECTEDLY REQUIRED

Normally, people do not need precision in their judgments. All that is needed is the combination of knowledge in the world and in the head that makes decisions unambiguous. Everything works just fine unless the environment changes so that the combined knowledge is no longer sufficient: this can lead to havoc. At least three countries discovered this fact the hard way: the United States, when it introduced the Susan B. Anthony one-dollar coin; Great Britain, a one-pound coin (before the switch to decimal currency); and France, a ten-franc coin (before the conversion to the common

European currency, the euro). The US dollar coin was confused with the existing twenty-five-cent piece (the quarter), and the British pound coin with the then five-pence piece that had the same diameter. Here is what happened in France:

> *PARIS With a good deal of fanfare, the French government released the new 10-franc coin (worth a little more than $1.50) on Oct. 22 [1986]. The public looked at it, weighed it, and began confusing it so quickly with the half-franc coin (worth only 8 cents) that a crescendo of fury and ridicule fell on both the government and the coin.*
>
> *Five weeks later, Minister of Finance Edouard Balladur suspended circulation of the coin. Within another four weeks, he canceled it altogether.*
>
> *In retrospect, the French decision seems so foolish that it is hard to fathom how it could have been made. After much study, designers came up with a silver-colored coin made of nickel and featuring a modernistic drawing by artist Joaquim Jimenez of a Gallic rooster on one side and of Marianne, the female symbol of the French republic, on the other. The coin was light, sported special ridges on its rim for easy reading by electronic vending machines and seemed tough to counterfeit.*
>
> *But the designers and bureaucrats were obviously so excited by their creation that they ignored or refused to accept the new coin's similarity to the hundreds of millions of silver-colored, nickel-based half-franc coins in circulation [whose] size and weight were perilously similar.* (Stanley Meisler. Copyright © 1986, *Los Angeles Times*. Reprinted with permission.)

The confusions probably occurred because the users of coins had already formed representations in their memories that were only sufficiently precise to distinguish among the coins that they were accustomed to using. Psychological research suggests that people maintain only partial descriptions of the things to be remembered. In the three examples of new coins introduced in the United States, Great Britain, and France, the descriptions formed to distinguish among national currency were not precise enough to distinguish between a new coin and at least one of the old coins.

Suppose I keep all my notes in a small red notebook. If this is my only notebook, I can describe it simply as "my notebook." If I buy several more notebooks, the earlier description will no longer work. Now I must identify the first one as small or red, or maybe both small and red, whichever allows me to distinguish it from the others. But what if I acquire several small red notebooks? Now I must find some other means of describing the first book, adding to the richness of the description and to its ability to discriminate among the several similar items. Descriptions need discriminate only among the choices in front of me, but what works for one purpose may not for another.

Not all similar-looking items cause confusion. In updating this edition of the book, I searched to see whether there might be more recent examples of coin confusions. I found this interesting item on the website Wikicoins.com:

> *Someday, a leading psychologist may weigh in on one of the perplexing questions of our time: if the American public was constantly confusing the Susan B. Anthony dollar with the roughly similar-sized quarter, how come they weren't also constantly confusing the $20 bill with the identical-sized $1 bill?* (James A. Capp, "Susan B. Anthony Dollar," at www.wikicoins.com. Retrieved May 29, 2012)

Here is the answer. Why not any confusion? We learn to discriminate among things by looking for distinguishing features. In the United States, size is one major way of distinguishing among coins, but not among paper money. With paper money, all the bills are the same size, so Americans ignore size and look at the printed numbers and images. Hence, we often confuse similar-size American coins but only seldom confuse similar-size American bills. But people who come from a country that uses size and color of their paper money to distinguish among the amounts (for example, Great Britain or any country that uses the euro) have learned to use size and color to distinguish among paper money and therefore are invariably confused when dealing with bills from the United States.

More confirmatory evidence comes from the fact that although long-term residents of Britain complained that they confused the one-pound coin with the five-pence coin, newcomers (and children) did not have the same confusion. This is because the long-term residents were working with their original set of descriptions, which did not easily accommodate the distinctions between these two coins. Newcomers, however, started off with no preconceptions and therefore formed a set of descriptions to distinguish among all the coins; in this situation, the one-pound coin offered no particular problem. In the United States, the Susan B. Anthony dollar coin never became popular and is no longer being made, so the equivalent observations cannot be made.

What gets confused depends heavily upon history: the aspects that have allowed us to distinguish among the objects in the past. When the rules for discrimination change, people can become confused and make errors. With time, they will adjust and learn to discriminate just fine and may even forget the initial period of confusion. The problem is that in many circumstances, especially one as politically charged as the size, shape, and color of currency, the public's outrage prevents calm discussion and does not allow for any adjustment time.

Consider this as an example of design principles interacting with the messy practicality of the real world. What appears good in principle can sometimes fail when introduced to the world. Sometimes, bad products succeed and good products fail. The world is complex.

CONSTRAINTS SIMPLIFY MEMORY

Before widespread literacy, and especially before the advent of sound recording devices, performers traveled from village to village, reciting epic poems thousands of lines long. This tradition still exists in some societies. How do people memorize such voluminous amounts of material? Do some people have huge amounts of knowledge in their heads? Not really. It turns out that external constraints exert control over the permissible choice of words, thus dramatically reducing the memory load. One of the secrets comes from the powerful constraints of poetry.

Consider the constraints of rhyming. If you wish to rhyme one word with another, there are usually a lot of alternatives. But if you must have a word with a particular meaning to rhyme with another, the joint constraints of meaning and rhyme can cause a dramatic reduction in the number of possible candidates, sometimes reducing a large set to a single choice. Sometimes there are no candidates at all. This is why it is much easier to memorize poetry than to create poems. Poems come in many different forms, but all have formal restrictions on their construction. The ballads and tales told by the traveling storytellers used multiple poetic constraints, including rhyme, rhythm, meter, assonance, alliteration, and onomatopoeia, while also remaining consistent with the story being told.

Consider these two examples:

> *One. I am thinking of three words: one means "a mythical being," the second is "the name of a building material," and the third is "a unit of time." What words do I have in mind?*
>
> *Two. This time look for rhyming words. I am thinking of three words: one rhymes with "post," the second with "eel," and the third with "ear." What words am I thinking of?* (From Rubin & Wallace, 1989.)

In both examples, even though you might have found answers, they were not likely to be the same three that I had in mind. There simply are not enough constraints. But suppose I now tell you that the words I seek are the same in both tasks: What is a word that means a mythical being and rhymes with "post"? What word is the name of a building material and rhymes with "eel"? And what word is a unit of time and rhymes with "ear"? Now the task is easy: the joint specification of the words completely constrains the selection. When the psychologists David Rubin and Wanda Wallace studied these examples in their laboratory, people almost never got the correct meanings or rhymes for the first two tasks, but most people correctly answered, "ghost," "steel," and "year" in the combined task.

The classic study of memory for epic poetry was done by Albert Bates Lord. In the mid-1900s he traveled throughout the former

Yugoslavia (now a number of separate, independent countries) and found people who still followed the oral tradition. He demonstrated that the "singer of tales," the person who learns epic poems and goes from village to village reciting them, is really re-creating them, composing poetry on the fly in such a way that it obeys the rhythm, theme, story line, structure, and other characteristics of the poem. This is a prodigious feat, but it is not an example of rote memory.

The power of multiple constraints allows one singer to listen to another singer tell a lengthy tale once, and then after a delay of a few hours or a day, to recite "the same song, word for word, and line for line." In fact, as Lord points out, the original and new recitations are not the same word for word, but both teller and listener perceive them as the same, even when the second version was twice as long as the first. They are the same in the ways that matter to the listener: they tell the same story, express the same ideas, and follow the same rhyme and meter. They are the same in all senses that matter to the culture. Lord shows just how the combination of memory for poetics, theme, and style combines with cultural structures into what he calls a "formula" for producing a poem perceived as identical to earlier recitations.

The notion that someone should be able to recite word for word is relatively modern. Such a notion can be held only after printed texts become available; otherwise who could judge the accuracy of a recitation? Perhaps more important, who would care?

All this is not to detract from the feat. Learning and reciting an epic poem, such as Homer's *Odyssey* and *Iliad,* is clearly difficult even if the singer is re-creating it: there are twenty-seven thousand lines of verse in the combined written version. Lord points out that this length is excessive, probably produced only during the special circumstances in which Homer (or some other singer) dictated the story slowly and repetitively to the person who first wrote it down. Normally the length would be varied to accommodate the whims of the audience, and no normal audience could sit through twenty-seven thousand lines. But even at one-third the size, nine thousand lines, being able to recite the poem is impressive: at one second per line, the verses would take two and one-half hours to recite. It is impressive even allowing for the fact that the poem is re-created as opposed to memorized, because neither the singer nor the audience expect word-for-word accuracy (nor would either have any way of verifying that).

Most of us do not learn epic poems. But we do make use of strong constraints that serve to simplify what must be retained in memory. Consider an example from a completely different domain: taking apart and reassembling a mechanical device. Typical items in the home that an adventuresome person might attempt to repair include a door lock, toaster, and washing machine. The device is apt to have tens of parts. What has to be remembered to be able to put the parts together again in a proper order? Not as much as might appear from an initial analysis. In the extreme case, if there are ten parts, there are 10! (ten factorial) different ways in which to reassemble them—a little over 3.5 million alternatives.

But few of these possibilities are possible: there are numerous physical constraints on the ordering. Some pieces must be assembled before it is even possible to assemble the others. Some pieces are physically constrained from fitting into the spots reserved for others: bolts must fit into holes of an appropriate diameter and depth; nuts and washers must be paired with bolts and screws of appropriate sizes; and washers must always be put on before nuts. There are even cultural constraints: we turn screws clockwise to tighten, counterclockwise to loosen; the heads of screws tend to go on the visible part (front or top) of a piece, bolts on the less visible part (bottom, side, or interior); wood screws and machine screws look different and are inserted into different kinds of materials. In the end, the apparently large number of decisions is reduced to only a few choices that should have been learned or otherwise noted during the disassembly. The constraints by themselves are often not sufficient to determine the proper reassembly of the device—mistakes do get made—but the constraints reduce the amount that must be learned to a reasonable quantity. Constraints are powerful tools for the designer: they are examined in detail in Chapter 4.

Memory Is Knowledge in the Head

An old Arabic folk tale, "'Ali Baba and the Forty Thieves," tells how the poor woodcutter 'Ali Baba discovered the secret cave of a band of thieves. 'Ali Baba overheard the thieves entering the cave and learned the secret phrase that opened the cave: "Open Simsim." (*Simsim* means "sesame" in Persian, so many versions of the story translate the phrase as "Open Sesame.") 'Ali Baba's brother-in-law, Kasim, forced him to reveal the secret. Kasim then went to the cave.

> *When he reached the entrance of the cavern, he pronounced the words, Open Simsim!*
>
> *The door immediately opened, and when he was in, closed on him. In examining the cave he was greatly astonished to find much more riches than he had expected from 'Ali Baba's relation.*
>
> *He quickly laid at the door of the cavern as many bags of gold as his ten mules could carry, but his thoughts were now so full of the great riches he should possess, that he could not think of the necessary words to make the door open. Instead of Open Simsim! he said Open Barley! and was much amazed to find that the door remained shut. He named several sorts of grain, but still the door would not open.*
>
> *Kasim never expected such an incident, and was so alarmed at the danger he was in that the more he endeavoured to remember the word Simsim the more his memory was confounded, and he had as much forgotten it as if he had never heard it mentioned.*
>
> *Kasim never got out. The thieves returned, cut off Kasim's head, and quartered his body.* (From Colum's 1953 edition of *The Arabian Nights.*)

Most of us will not get our head cut off if we fail to remember a secret code, but it can still be very hard to recall the code. It is one thing to have to memorize one or two secrets: a combination, or a password, or the secret to opening a door. But when the number of secret codes gets too large, memory fails. There seems to be a conspiracy, one calculated to destroy our sanity by overloading our memory. Many codes, such as postal codes and telephone numbers, exist primarily to make life easier for machines and their designers without any consideration of the burden placed upon people. Fortunately, technology has now permitted most of us to avoid having to remember this arbitrary knowledge but to let our technology do it for us: phone numbers, addresses and postal codes, Internet and e-mail addresses are all retrievable automatically, so we no longer have to learn them. Security codes, however, are a different matter, and in the never-ending, escalating battle between the white hats and the black, the good guys and the bad, the number of different arbitrary codes we must remember or special security devices we must carry with us continues to escalate in both number and complexity.

Many of these codes must be kept secret. There is no way that we can learn all those numbers or phrases. Quick: what magical command was Kasim trying to remember to open the cavern door?

How do most people cope? They use simple passwords. Studies show that five of the most common passwords are: "password," "123456," "12345678," "qwerty," and "abc123." All of these are clearly selected for easy remembering and typing. All are therefore easy for a thief or mischief-maker to try. Most people (including me) have a small number of passwords that they use on as many different sites as possible. Even security professionals admit to this, thereby hypocritically violating their own rules.

Many of the security requirements are unnecessary, and needlessly complex. So why are they required? There are many reasons. One is that there are real problems: criminals impersonate identities to steal people's money and possessions. People invade others' privacy, for nefarious or even harmless purposes. Professors and teachers need to safeguard examination questions and grades. For companies and nations, it is important to maintain secrets. There are lots of reasons to keep things behind locked doors or password-protected walls. The problem, however, is the lack of proper understanding of human abilities.

We do need protection, but most of the people who enforce the security requirements at schools, businesses, and government are technologists or possibly law-enforcement officials. They understand crime, but not human behavior. They believe that "strong" passwords, ones difficult to guess, are required, and that they must be changed frequently. They do not seem to recognize that we now need so many passwords—even easy ones—that it is difficult to remember which goes with which requirement. This creates a new layer of vulnerability.

The more complex the password requirements, the less secure the system. Why? Because people, unable to remember all these combinations, write them down. And then where do they store this private, valuable knowledge? In their wallet, or taped under the computer keyboard, or wherever it is easy to find, because it is so frequently needed. So a thief only has to steal the wallet or find the list and then all secrets are known. Most people are honest, concerned workers. And it is these individuals that complex security systems impede the most, preventing them from getting their work done. As a result, it is often the most dedicated employee who violates the security rules and weakens the overall system.

When I was doing the research for this chapter, I found numerous examples of secure passwords that force people to use insecure memory devices for them. One post on the "Mail Online" forum of the British *Daily Mail* newspaper described the technique:

> *When I used to work for the local government organisation we HAD TO change our Passwords every three months. To ensure I could remember it, I used to write it on a Post-It note and stick it above my desk.*

How can we remember all these secret things? Most of us can't, even with the use of mnemonics to make some sense of nonsensical material. Books and courses on improving memory can work, but the methods are laborious to learn and need continual practice to maintain. So we put the memory in the world, writing things down in books, on scraps of paper, even on the backs of our hands. But we disguise them to thwart would-be thieves. That creates another problem: How do we disguise the items, how do we hide them, and how do we remember what the disguise was or where we put it? Ah, the foibles of memory.

Where should you hide something so that nobody else will find it? In unlikely places, right? Money is hidden in the freezer; jewelry in the medicine cabinet or in shoes in the closet. The key to the front door is hidden under the mat or just below the window ledge. The car key is under the bumper. The love letters are in a flower vase. The problem is, there aren't that many unlikely places in the home. You may not remember where the love letters or keys are hidden, but your burglar will. Two psychologists who examined the issue described the problem this way:

> *There is often a logic involved in the choice of unlikely places. For example, a friend of ours was required by her insurance company to acquire a safe if she wished to insure her valuable gems. Recognizing that she might forget the combination to the safe, she thought carefully about where to keep the combination. Her solution was to write it in her personal phone directory under the letter S next to "Mr. and Mrs. Safe," as if it were a telephone number. There is a clear logic here: Store numerical information with other numerical information. She was appalled, however, when she heard a reformed burglar on a daytime television talk show say that upon encountering a safe, he always headed for the phone directory because many people keep the combination there.* (From Winograd & Soloway, 1986, "On Forgetting the Locations of Things Stored in Special Places." Reprinted with permission.)

All the arbitrary things we need to remember add up to unwitting tyranny. It is time for a revolt. But before we revolt, it is important to know the solution. As noted earlier, one of my self-imposed rules is, "Never criticize unless you have a better alternative." In this case, it is not clear what the better system might be.

Some things can only be solved by massive cultural changes, which probably means they will never be solved. For example, take the problem of identifying people by their names. People's names evolved over many thousands of years, originally simply to distinguish people within families and groups who lived together. The use of multiple names (given names and surnames) is relatively recent, and even those do not distinguish one person from all the seven billion in the world. Do we write the given name first, or the surname? It depends upon what country you are in. How many names does a person have? How many characters in a name? What characters are legitimate? For example, can a name include a digit? (I know people who have tried to use such names as "h3nry." I know of a company named "Autonom3.")

How does a name translate from one alphabet to another? Some of my Korean friends have given names that are identical when written in the Korean alphabet, Hangul, but that are different when transliterated into English.

Many people change their names when they get married or divorced, and in some cultures, when they pass significant life events. A quick search on the Internet reveals multiple questions from people in Asia who are confused about how to fill out American or European passport forms because their names don't correspond to the requirements.

And what happens when a thief steals a person's identity, masquerading as the other individual, using his or her money and credit? In the United States, these identity thieves can also apply for income tax rebates and get them, and when the legitimate taxpayers try to get their legitimate refund, they are told they already received it.

I once attended a meeting of security experts that was held at the corporate campus of Google. Google, like most corporations, is very protective of its processes and advanced research projects, so most of the buildings were locked and guarded. Attendees of the security meeting were not allowed access (except those who worked at Google, of course). Our meetings were held in a conference room in the public space of an otherwise secure building. But the toilets were all located inside a secure area. How did we manage? These world-famous, leading authorities on security figured out a solution: They found a brick and used it to prop open the door leading into the secure area. So much for security: Make something too secure, and it becomes less secure.

How do we solve these problems? How do we guarantee people's access to their own records, bank accounts, and computer systems? Almost any scheme you can imagine has already been proposed, studied, and found to have defects. Biometric markers (iris or retina patterns, fingerprints, voice recognition, body type, DNA)? All can be forged or the systems' databases manipulated. Once someone manages to fool the system, what recourse is there? It isn't possible to change biometric markers, so once they point to the wrong person, changes are extremely difficult to make.

The strength of a password is actually pretty irrelevant because most passwords are obtained through "key loggers" or are stolen. A key logger is software hidden within your computer system that records what you type and sends it to the bad guys. When computer systems are broken into, millions of passwords might get stolen, and even if they are encrypted, the bad guys can often decrypt them. In both these cases, however secure the password, the bad guys know what it is.

The safest methods require multiple identifiers, the most common schemes requiring at least two different kinds: "something you have" plus "something you know." The "something you have" is often a physical identifier, such as a card or key, perhaps even something implanted under the skin or a biometric identifier, such as fingerprints or patterns of the eye's iris. The "something you know" would be knowledge in the head, most likely something memorized. The memorized item doesn't have to be as secure as today's passwords because it wouldn't work without the "something you have." Some systems allow for a second, alerting password, so that if the bad guys try to force someone to enter a password into a system, the individual would use the alerting one, which would warn the authorities of an illegal entry.

Security poses major design issues, ones that involve complex technology as well as human behavior. There are deep, fundamental difficulties. Is there a solution? No, not yet. We will probably be stuck with these complexities for a long time.

The Structure of Memory

Say aloud the numbers 1, 7, 4, 2, 8. Next, without looking back, repeat them. Try again if you must, perhaps closing your eyes, the better to "hear" the sound still echoing in mental activity. Have someone read a random sentence to you. What were the words? The memory of the just present is available immediately, clear and complete, without mental effort.

What did you eat for dinner three days ago? Now the feeling is different. It takes time to recover the answer, which is neither as clear nor as complete a remembrance as that of the just present, and the recovery is likely to require considerable mental effort. Retrieval of the past differs from retrieval of the just present. More effort is required, less clarity results. Indeed, the "past" need not be so long ago. Without looking back, what were those digits? For some people, this retrieval now takes time and effort. (From *Learning and Memory*, Norman, 1982.)

Psychologists distinguish between two major classes of memory: short-term or working memory, and long-term memory. The two are quite different, with different implications for design.

SHORT-TERM OR WORKING MEMORY

Short-term or working memory (STM) retains the most recent experiences or material that is currently being thought about. It is the memory of the just present. Information is retained automatically and retrieved without effort; but the amount of information that can be retained this way is severely limited. Something like five to seven items is the limit of STM, with the number going to ten or twelve if the material is continually repeated, what psychologists call "rehearsing."

Multiply 27 times 293 in your head. If you try to do it the same way you would with paper and pencil, you will almost definitely be unable to hold all the digits and intervening answers within STM. You will fail. The traditional method of multiplying is optimized for paper and pencil. There is no need to minimize the burden on working memory because the numbers written on the paper serve this function (knowledge in the world), so the burden on STM, on knowledge in the head, is quite limited. There are ways of doing mental multiplication, but the methods are quite different from those using paper and pencil and require considerable training and practice.

Short-term memory is invaluable in the performance of everyday tasks, in letting us remember words, names, phrases, and parts of tasks: hence its alternative name, working memory. But the material being maintained in STM is quite fragile. Get distracted by some other activity and, poof, the stuff in STM disappears. It is capable of holding a postal code or telephone number from the time you look it up until the time it is used—as long as no distractions occur. Nine- or ten-digit numbers give trouble, and when the number starts to exceed that—don't bother. Write it down. Or divide the number into several shorter segments, transforming the long number into meaningful chunks.

Memory experts use special techniques, called *mnemonics,* to remember amazingly large amounts of material, often after only a single exposure. One method is to transform the digits into meaningful segments (one famous study showed how an athlete thought of digit sequences as running times, and after refining the method over a long period, could learn incredibly long sequences at one glance). One traditional method used to encode long sequences of digits is to first transform each digit into a consonant, then transform the consonant sequence into a memorable phrase. A standard table of conversions of digits to consonants has been around for hundreds of years, cleverly designed to be easy to learn because the consonants can be derived from the shape of the digits. Thus, "1" is translated into "t" (or the similar-sounding "d"), "2" becomes "n," "3" becomes "m," "4" is "r," and "5" becomes "L" (as in the Roman numeral for 50). The full table and the mnemonics for learning the pairings are readily found on the Internet by searching for "number-consonant mnemonic."

Using the number-consonant transformation, the string 4194780135092770 translates into the letters *rtbrkfstmlspncks,* which in turn may become, "A hearty breakfast meal has pancakes." Most people are not experts at retaining long arbitrary strings of anything, so although it is interesting to observe memory wizards, it would be wrong to design systems that assumed this level of proficiency.

The capacity of STM is surprisingly difficult to measure, because how much can be retained depends upon the familiarity of the material. Retention, moreover, seems to be of meaningful items, rather than of some simpler measure such as seconds or individual sounds or letters. Retention is affected by both time and the number of items. The number of items is more important than time, with each new item decreasing the likelihood of remembering all of the preceding items. The capacity is items because people can remember roughly the same number of digits and words, and almost the same number of simple three- to five-word phrases. How can this be? I suspect that STM holds something akin to a pointer to an already encoded item in long-term memory, which means the memory capacity is the number of pointers it can keep. This would account for the fact that the length or complexity of the item has little impact—simply the number of items. It doesn't neatly account for the fact that we make acoustical errors in STM, unless the pointers are held in a kind of acoustical memory. This remains an open topic for scientific exploration.

The traditional measures of STM capacity range from five to seven, but from a practical point of view, it is best to think of it as holding only three to five items. Does that seem too small a number? Well, when you meet a new person, do you always remember his or her name? When you have to dial a phone number, do you have to look at it several times while entering the digits? Even minor distractions can wipe out the stuff we are trying to hold on to in STM.

What are the design implications? Don't count on much being retained in STM. Computer systems often enhance people's frustration when things go wrong by presenting critical information in a message that then disappears from the display just when the person wishes to make use of the information. So how can people remember the critical information? I am not surprised when people hit, kick, or otherwise attack their computers.

I have seen nurses write down critical medical information about their patients on their hands because the critical information would disappear if the nurse was distracted for a moment by someone asking a question. The electronic medical records systems automatically log out users when the system does not appear to be in use. Why the automatic logouts? To protect patient privacy. The cause may be well motivated, but the action poses severe challenges to nurses who are continually being interrupted in their work by physicians, co-workers, or patient requests. While they are attending to the interruption, the system logs them out, so they have to start over again. No wonder these nurses wrote down the knowledge, although this then negated much of the value of the computer system in minimizing handwriting errors. But what else were they to do? How else to get at the critical information? They couldn't remember it all: that's why they had computers.

The limits on our short-term memory systems caused by interfering tasks can be mitigated by several techniques. One is through the use of multiple sensory modalities. Visual information does not much interfere with auditory, actions do not interfere much with either auditory or written material. Haptics (touch) is also minimally interfering. To maximize efficiency of working memory it is best to present different information over different modalities: sight, sound, touch (haptics), hearing, spatial location, and gestures. Automobiles should use auditory presentation of driving instructions and haptic vibration of the appropriate side of the driver's seat or steering wheel to warn when drivers leave their lanes, or when there are other vehicles to the left or right, so as not to interfere with the visual processing of driving information. Driving is primarily visual, so the use of auditory and haptic modalities minimizes interference with the visual task.

LONG-TERM MEMORY

Long-term memory (LTM) is memory for the past. As a rule, it takes time for information to get into LTM and time and effort to get it out again. Sleep seems to play an important role in strengthening the memories of each day's experiences. Note that we do not remember our experiences as an exact recording; rather, as bits and pieces that are reconstructed and interpreted each time we recover the memories, which means they are subject to all the distortions and changes that the human explanatory mechanism imposes upon life. How well we can ever recover experiences and knowledge from LTM is highly dependent upon how the material was interpreted in the first place. What is stored in LTM under one interpretation probably cannot be found later on when sought under some other interpretation. As for how large the memory is, nobody really knows: giga- or tera-items. We don't even know what kinds of units should be used. Whatever the size, it is so large as not to impose any practical limit.

The role of sleep in the strengthening of LTM is still not well understood, but there are numerous papers investigating the topic. One possible mechanism is that of rehearsal. It has long been known that rehearsal of material—mentally reviewing it while still active in working memory (STM)—is an important component of the formation of long-term memory traces. "Whatever makes you rehearse during sleep is going to determine what you remember later, and conversely, what you're going to forget," said Professor Ken Paller of Northwestern University, one of the authors of a recent study on the topic (Oudiette, Antony, Creery, and Paller, 2013). But although rehearsal in sleep strengthens memories, it might also falsify them: "Memories in our brain are changing all of the time. Sometimes you improve memory storage by rehearsing all the details, so maybe later you remember better—or maybe worse if you've embellished too much."

Remember how you answered this question from Chapter 2?

> *In the house you lived in three houses ago, as you entered the front door, was the doorknob on the left or right?*

For most people, the question requires considerable effort just to recall which house is involved, plus one of the special techniques described in Chapter 2 for putting yourself back at the scene and reconstructing the answer. This is an example of procedural memory, a memory for how we do things, as opposed to declarative memory, the memory for factual information. In both cases, it can take considerable time and effort to get to the answer. Moreover, the answer is not directly retrieved in a manner analogous to the way we read answers from books or websites. The answer is a reconstruction of the knowledge, so it is subject to biases and distortions. Knowledge in memory is meaningful, and at the time of retrieval, a person might subject it to a different meaningful interpretation than is wholly accurate.

A major difficulty with LTM is in organization. How do we find the things we are trying to remember? Most people have had the "tip of the tongue" experience when trying to remember a name or word: there is a feeling of knowing, but the knowledge is not consciously available. Sometime later, when engaged in some other, different activity, the name may suddenly pop into the conscious mind. The way by which people retrieve the needed knowledge is still unknown, but probably involves some form of pattern-matching mechanism coupled with a confirmatory process that checks for consistency with the required knowledge. This is why when you search for a name but continually retrieve the wrong name, you know it is wrong. Because this false retrieval impedes the correct retrieval, you have to turn to some other activity to allow the subconscious memory retrieval process to reset itself.

Because retrieval is a reconstructive process, it can be erroneous. We may reconstruct events the way we would prefer to remember them, rather than the way we experienced them. It is relatively easy to bias people so that they form false memories, "remembering" events in their lives with great clarity, even though they never occurred. This is one reason that eyewitness testimony in courts of law is so problematic: eyewitnesses are notoriously unreliable. A huge number of psychological experiments show how easy it is to implant false memories into people's minds so convincingly that people refuse to admit that the memory is of an event that never happened.

Knowledge in the head is actually knowledge in memory: internal knowledge. If we examine how people use their memories and how they retrieve knowledge, we discover a number of categories. Two are important for us now:

1. **Memory for arbitrary things.** The items to be retained seem arbitrary, with no meaning and no particular relationship to one another or to things already known.
2. **Memory for meaningful things.** The items to be retained form meaningful relationships with themselves or with other things already known.

MEMORY FOR ARBITRARY AND MEANINGFUL THINGS

Arbitrary knowledge can be classified as the simple remembering of things that have no underlying meaning or structure. A good example is the memory of the letters of the alphabet and their ordering, the names of people, and foreign vocabulary, where there appears to be no obvious structure to the material. This also applies to the learning of the arbitrary key sequences, commands, gestures, and procedures of much of our modern technology: This is rote learning, the bane of modern existence.

Some things do require rote learning: the letters of the alphabet, for example, but even here we add structure to the otherwise meaningless list of words, turning the alphabet into a song, using the natural constraints of rhyme and rhythm to create some structure.

Rote learning creates problems. First, because what is being learned is arbitrary, the learning is difficult: it can take considerable time and effort. Second, when a problem arises, the memorized sequence of actions gives no hint of what has gone wrong, no suggestion of what might be done to fix the problem. Although some things are appropriate to learn by rote, most are not. Alas, it is still the dominant method of instruction in many school systems, and even for much adult training. This is how some people are taught to use computers, or to cook. It is how we have to learn to use some of the new (poorly designed) gadgets of our technology.

We learn arbitrary associations or sequences by artificially providing structure. Most books and courses on methods for improving memory (mnemonics) use a variety of standard methods for providing structure, even for things that might appear completely arbitrary, such as grocery lists, or matching the names of people to their appearance. As we saw in the discussion of these methods for STM, even strings of digits can be remembered if they can be associated with meaningful structures. People who have not received this training or who have not invented some methods themselves often try to manufacture some artificial structure, but these are often rather unsatisfactory, which is why the learning is so bad.

Most things in the world have a sensible structure, which tremendously simplifies the memory task. When things make sense, they correspond to knowledge that we already have, so the new material can be understood, interpreted, and integrated with previously acquired material. Now we can use rules and constraints to help understand what things go together. Meaningful structure can organize apparent chaos and arbitrariness.

Remember the discussion of conceptual models in Chapter 1? Part of the power of a good conceptual model lies in its ability to provide meaning to things. Let's look at an example to show how a meaningful interpretation transforms an apparently arbitrary task into a natural one. Note that the appropriate interpretation may not at first be obvious; it, too, is knowledge and has to be discovered.

A Japanese colleague, Professor Yutaka Sayeki of the University of Tokyo, had difficulty remembering how to use the turn signal switch on his motorcycle's left handlebar. Moving the switch forward signaled a right turn; backward, a left turn. The meaning of the switch was clear and unambiguous, but the direction in which it should be moved was not. Sayeki kept thinking that because the switch was on the left handlebar, pushing it forward should signal a left turn. That is, he was trying to map the action "push the left switch forward" to the intention "turn left," which was wrong. As a result, he had trouble remembering which switch direction should be used for which turning direction. Most motorcycles have the turn-signal switch mounted differently, rotated 90 degrees, so that moving it left signals a left turn; moving it right, a right turn. This mapping is easy to learn (it is an example of a natural mapping, discussed at the end of this chapter). But the turn switch on Sayeki's motorcycle moved forward and back, not left and right. How could he learn it?

Sayeki solved the problem by reinterpreting the action. Consider the way the handlebars of the motorcycle turn. For a left turn, the left handlebar moves backward. For a right turn, the left handlebar moves forward. The required switch movements exactly paralleled the handlebar movements. If the task is conceptualized as signaling the direction of motion of the handlebars rather than the direction of the motorcycle, the switch motion can be seen to mimic the desired motion; finally we have a natural mapping.

When the motion of the switch seemed arbitrary, it was difficult to remember. Once Professor Sayeki had invented a meaningful relationship, he found it easy to remember the proper switch operation. (Experienced riders will point out that this conceptual model is wrong: to turn a bike, one first steers in the opposite direction of the turn. This is discussed as Example 3 in the next section, "Approximate Models.")

The design implications are clear: provide meaningful structures. Perhaps a better way is to make memory unnecessary: put the required information in the world. This is the power of the traditional graphical user interface with its old-fashioned menu structure. When in doubt, one could always examine all the menu items until the desired one was found. Even systems that do not use menus need to provide some structure: appropriate constraints and forcing functions, natural good mapping, and all the tools of feedforward and feedback. The most effective way of helping people remember is to make it unnecessary.

Approximate Models: Memory in the Real World

Conscious thinking takes time and mental resources. Well-learned skills bypass the need for conscious oversight and control: conscious control is only required for initial learning and for dealing with unexpected situations. Continual practice automates the action cycle, minimizing the amount of conscious thinking and problem-solving required to act. Most expert, skilled behavior works this way, whether it is playing tennis or a musical instrument, or doing mathematics and science. Experts minimize the need for conscious reasoning. Philosopher and mathematician Alfred North Whitehead stated this principle over a century ago:

> *It is a profoundly erroneous truism, repeated by all copy-books and by eminent people when they are making speeches, that we should cultivate the habit of thinking of what we are doing. The precise opposite is the case. Civilization advances by extending the number of important operations which we can perform without thinking about them.* (Alfred North Whitehead, 1911.)

One way to simplify thought is to use simplified models, approximations to the true underlying state of affairs. Science deals in truth, practice deals with approximations. Practitioners don't need truth: they need results relatively quickly that, although inaccurate, are "good enough" for the purpose to which they will be applied. Consider these examples:

EXAMPLE 1: CONVERTING TEMPERATURES BETWEEN FAHRENHEIT AND CELSIUS

It is now 55°F outside my home in California. What temperature is it in Celsius? Quick, do it in your head without using any technology: What is the answer?

I am sure all of you remember the conversion equation:

$$°C = (°F–32) \times 5 / 9$$

Plug in 55 for °F, and °C = (55–32) × 5 / 9 = 12.8°. But most people can't do this without pencil and paper because there are too many intermediate numbers to maintain in STM.

Want a simpler way? Try this approximation—you can do it in your head, there is no need for paper or pencil:

$$°C = (°F–30) / 2$$

Plug in 55 for °F, and °C = (55–30) / 2 = 12.5°. Is the equation an exact conversion? No, but the approximate answer of 12.5 is close enough to the correct value of 12.8. After all, I simply wanted to know whether I should wear a sweater. Anything within 5°F of the real value would work for this purpose.

Approximate answers are often good enough, even if technically wrong. This simple approximation method for temperature conversion is "good enough" for temperatures in the normal range of interior and outside temperatures: it is within 3°F (or 1.7°C) in the range of –5° to 25°C (20° to 80°F). It gets further off at lower or higher temperatures, but for everyday use, it is wonderful. Approximations are good enough for practical use.

EXAMPLE 2: A MODEL OF SHORT-TERM MEMORY

Here is an approximate model for STM:

> *There are five memory slots in short-term memory. Each time a new item is added, it occupies a slot, knocking out whatever was there beforehand.*

Is this model true? No, not a single memory researcher in the entire world believes this to be an accurate model of STM. But it is good enough for applications. Make use of this model, and your designs will be more usable.

EXAMPLE 3: STEERING A MOTORCYCLE

In the preceding section, we learned how Professor Sayeki mapped the turning directions of his motorcycle to his turn signals, enabling him to remember their correct usage. But there, I also pointed out that the conceptual model was wrong.

Why is the conceptual model for steering a motorcycle useful even though it is wrong? Steering a motorcycle is counterintuitive: to turn to the left, the handlebars must first be turned to the right. This is called countersteering, and it violates most people's conceptual models. Why is this true? Shouldn't we rotate the handlebars left to turn the bike left? The most important component of turning a two-wheeled vehicle is lean: when the bike is turning left, the rider is leaning to the left. Countersteering causes the rider to lean properly: when the handlebars are turned to the right, the resulting forces upon the rider cause the body to lean left. This weight shift then causes the bike to turn left.

Experienced riders often do the correct operations subconsciously, unaware that they start a turn by rotating the handlebars opposite from the intended direction, thus violating their own conceptual models. Motorcycle training courses have to conduct special exercises to convince riders that this is what they are doing.

You can test this counterintuitive concept on a bicycle or motorcycle by getting up to a comfortable speed, placing the palm of the hand on the end of the left handlebar, and gently pushing it forward. The handlebars and front wheel will turn to the right and the body will lean to the left, resulting in the bike—and the handlebars—turning to the left.

Professor Sayeki was fully aware of this contradiction between his mental scheme and reality, but he wanted his memory aid to match his conceptual model. Conceptual models are powerful explanatory devices, useful in a variety of circumstances. They do not have to be accurate as long as they lead to the correct behavior in the desired situation.

EXAMPLE 4: "GOOD ENOUGH" ARITHMETIC

Most of us can't multiply two large numbers in our head: we forget where we are along the way. Memory experts can multiply two large numbers quickly and effortlessly in their heads, amazing audiences with their skills. Moreover, the numbers come out left to right, the way we use them, not right to left, as we write them while laboriously using pencil and paper to compute the answers. These experts use special techniques that minimize the load on working memory, but they do so at the cost of having to learn numerous special methods for different ranges and forms of problems.

Isn't this something we should all learn? Why aren't school systems teaching this? My answer is simple: Why bother? I can estimate the answer in my head with reasonable accuracy, often good enough for the purpose. When I need precision and accuracy, well, that's what calculators are for.

Remember my earlier example, to multiply 27 times 293 in your head? Why would anyone need to know the precise answer? an approximate answer is good enough, and pretty easy to get. Change 27 to 30, and 293 to 300: 30 × 300 = 9,000 (3 × 3 = 9, and add back the three zeros). The accurate answer is 7,911, so the estimate of 9,000 is only 14 percent too large. In many instances, this is good enough. Want a bit more accuracy? We changed 27 to 30 to make the multiplication easier. That's 3 too large. So subtract 3 × 300 from the answer (9,000 – 900). Now we get 8,100, which is accurate within 2 percent.

It is rare that we need to know the answers to complex arithmetic problems with great precision: almost always, a rough estimate is good enough. When precision is required, use a calculator. That's what machines are good for: providing great precision. For most purposes, estimates are good enough. Machines should focus on solving arithmetic problems. People should focus on higher-level issues, such as the reason the answer was needed.

Unless it is your ambition to become a nightclub performer and amaze people with great skills of memory, here is a simpler way to dramatically enhance both memory and accuracy: write things down. Writing is a powerful technology: why not use it? Use a pad of paper, or the back of your hand. Write it or type it. Use a phone or a computer. Dictate it. This is what technology is for.

The unaided mind is surprisingly limited. It is things that make us smart. Take advantage of them.

SCIENTIFIC THEORY VERSUS EVERYDAY PRACTICE

Science strives for truth. As a result, scientists are always debating, arguing, and disagreeing with one another. The scientific method is one of debate and conflict. Only ideas that have passed through the critical examination of multiple other scientists survive. This continual disagreement often seems strange to the nonscientist, for it appears that scientists don't know anything. Select almost any topic, and you will discover that scientists who work in that area are continually disagreeing.

But the disagreements are illusory. That is, most scientists usually agree about the broad details: their disagreements are often about tiny details that are important for distinguishing between two competing theories, but that might have very little impact in the real world of practice and applications.

In the real, practical world, we don't need absolute truth: approximate models work just fine. Professor Sayeki's simplified conceptual model of steering his motorcycle enabled him to remember which way to move the switches for his turn signals; the simplified equation for temperature conversion and the simplified model of approximate arithmetic enabled "good enough" answers in the head. The simplified model of STM provides useful design guidance, even if it is scientifically wrong. Each of these approximations is wrong, yet all are valuable in minimizing thought, resulting in quick, easy results whose accuracy is "good enough."

Knowledge in the Head

Knowledge in the world, external knowledge, is a valuable tool for remembering, but only if it is available at the right place, at the right time, in the appropriate situation. Otherwise, we must use knowledge in the head, in the mind. A folk saying captures this situation well: "Out of sight, out of mind." Effective memory uses all the clues available: knowledge in the world and in the head, combining world and mind. We have already seen how the combination allows us to function quite well in the world even though either source of knowledge, by itself, is insufficient.

HOW PILOTS REMEMBER WHAT AIR-TRAFFIC CONTROL TELLS THEM

Airplane pilots have to listen to commands from air-traffic control delivered at a rapid pace, and then respond accurately. Their lives depend upon being able to follow the instructions accurately. One website, discussing the problem, gave this example of instructions to a pilot about to take off for a flight:

Frasca 141, cleared to Mesquite airport, via turn left heading 090, radar vectors to Mesquite airport. Climb and maintain 2,000. Expect 3,000 10 minutes after departure. Departure frequency 124.3, squawk 5270.
(Typical Air traffic control sequence, usually spoken extremely rapidly. Text from "ATC Phraseology," on numerous websites, with no credit for originator.)

"How can we remember all that," asked one novice pilot, "when we are trying to focus on taking off?" Good question. Taking off is a busy, dangerous procedure with a lot going on, both inside and outside the airplane. How do pilots remember? Do they have superior memories?

Pilots use three major techniques:

1. They write down the critical information.
2. They enter it into their equipment as it is told to them, so minimal memory is required.
3. They remember some of it as meaningful phrases.

Although to the outside observer, all the instructions and numbers seem random and confusing, to the pilots they are familiar names, familiar numbers. As one respondent pointed out, those are common numbers and a familiar pattern for a takeoff. "Frasca 141" is the name of the airplane, announcing the intended recipient of these instructions. The first critical item to remember is to turn left to a compass direction of 090, then climb to an altitude of 2,000 feet. Write those two numbers down. Enter the radio frequency 124.3 into the radio as you hear it—but most of the time this frequency is known in advance, so the radio is probably already set to it. All you have to do is look at it and see that it is set properly. Similarly, setting the "squawk box to 5270" is the special code the airplane sends whenever it is hit by a radar signal, identifying the airplane to the air-traffic controllers. Write it down, or set it into the equipment as it is being said. As for the one remaining item, "Expect 3,000 10 minutes after departure," nothing need be done. This is just reassurance that in ten minutes, Frasca 141 will probably be advised to climb to 3,000 feet, but if so, there will be a new command to do so.

How do pilots remember? They transform the new knowledge they have just received into memory in the world, sometimes by writing, sometimes by using the airplane's equipment.

The design implication? The easier it is to enter the information into the relevant equipment as it is heard, the less chance of memory error. The air-traffic control system is evolving to help. The instructions from the air-traffic controllers will be sent digitally, so that they can remain displayed on a screen as long as the pilot wishes. The digital transmission also makes it easy for automated equipment to set itself to the correct parameters. Digital transmission of the controller's commands has some disadvantages, however. Other aircraft will not hear the commands, which reduces pilot awareness of what all the airplanes in the vicinity are going to do. Researchers in air-traffic control and aviation safety are looking into these issues. Yes, it's a design issue.

REMINDING: PROSPECTIVE MEMORY

The phrases *prospective memory* or *memory for the future* might sound counterintuitive, or perhaps like the title of a science-fiction novel, but to memory researchers, the first phrase simply denotes the task of remembering to do some activity at a future time. The second phrase denotes planning abilities, the ability to imagine future scenarios. Both are closely related.

Consider reminding. Suppose you have promised to meet some friends at a local café on Wednesday at three thirty in the afternoon. The knowledge is in your head, but how are you going to remember it at the proper time? You need to be reminded. This is a clear instance of prospective memory, but your ability to provide the required cues involves some aspect of memory for the future as well. Where will you be Wednesday just before the planned meeting? What can you think of now that will help you remember then?

There are many strategies for reminding. One is simply to keep the knowledge in your head, trusting yourself to recall it at the critical time. If the event is important enough, you will have no problem remembering it. It would be quite strange to have to set a calendar alert to remind yourself, "Getting married at 3 PM."

Relying upon memory in the head is not a good technique for commonplace events. Ever forget a meeting with friends? It happens a lot. Not only that, but even if you might remember the appointment, will you remember all the details, such as that you intended to loan a book to one of them? Going shopping, you may remember to stop at the store on the way home, but will you remember all the items you were supposed to buy?

If the event is not personally important and several days away, it is wise to transfer some of the burden to the world: notes, calendar reminders, special cell phone or computer reminding services. You can ask friends to remind you. Those of us with assistants put the burden on them. They, in turn, write notes, enter events on calendars, or set alarms on their computer systems.

Why burden other people when we can put the burden on the thing itself? Do I want to remember to take a book to a colleague? I put the book someplace where I cannot fail to see it when I leave the house. A good spot is against the front door so that I can't leave without tripping over it. Or I can put my car keys on it, so when I leave, I am reminded. Even if I forget, I can't drive away without the keys. (Better yet, put the keys under the book, else I might still forget the book.)

There are two different aspects to a reminder: the signal and the message. Just as in doing an action we can distinguish between knowing *what* can be done and knowing *how* to do it, in reminding we must distinguish between the *signal*—knowing that something is to be remembered, and the *message*—remembering the information itself. Most popular reminding methods typically provide only one or the other of these two critical aspects. The famous "tie a string around your finger" reminder provides only the signal. It gives no hint of what is to be remembered. Writing a note to yourself provides only the message; it doesn't remind you ever to look at it. The ideal reminder has to have both components: the signal that something is to be remembered, and then the message of what it is.

The signal that something is to be remembered can be a sufficient memory cue if it occurs at the correct time and place. Being reminded too early or too late is just as useless as having no reminder. But if the reminder comes at the correct time or location, the environmental cue can suffice to provide enough knowledge to aid retrieval of the to-be-remembered item. Time-based reminders can be effective: the *bing* of my cell phone reminds me of the next appointment. Location-based reminders can be effective in giving the cue at the precise place where it will be needed. All the knowledge needed can reside in the world, in our technology.

The need for timely reminders has created loads of products that make it easier to put the knowledge in the world—timers, diaries, calendars. The need for electronic reminders is well known, as the proliferation of apps for smart phones, tablets, and other portable devices attests. Yet surprisingly in this era of screen-based devices, paper tools are still enormously popular and effective, as the number of paper-based diaries and reminders indicates.

The sheer number of different reminder methods also indicates that there is indeed a great need for assistance in remembering, but that none of the many schemes and devices is completely satisfactory. After all, if any one of them was, then we wouldn't need so many. The less effective ones would disappear and new schemes would not continually be invented.

The Tradeoff Between Knowledge in the World and in the Head

Knowledge in the world and knowledge in the head are both essential in our daily functioning. But to some extent we can choose to lean more heavily on one or the other. That choice requires a tradeoff—gaining the advantages of knowledge in the world means losing the advantages of knowledge in the head (Table 3.1).

Knowledge in the world acts as its own reminder. It can help us recover structures that we otherwise would forget. Knowledge in the head is efficient: no search and interpretation of the environment is required. The tradeoff is that to use our knowledge in the head, we have to be able to store and retrieve it, which might require considerable amounts of learning. Knowledge in the world requires no learning, but can be more difficult to use. And it relies heavily upon the continued physical presence of the knowledge; change the environment and the knowledge might be lost. Performance relies upon the physical stability of the task environment.

As we just discussed, reminders provide a good example of the relative tradeoffs between knowledge in the world versus in the head. Knowledge in the world is accessible. It is self-reminding. It is always there, waiting to be seen, waiting to be used. That is why we structure our offices and our places of work so carefully. We put piles of papers where they can be seen, or if we like a clean desk, we put them in standardized locations and teach ourselves (knowledge in the head) to look in these standard places routinely. We use clocks and calendars and notes. Knowledge in the mind

TABLE 3.1. **Tradeoffs Between Knowledge in the World and in the Head**

Knowledge in the World	**Knowledge in the Head**
Information is readily and easily available whenever perceivable.	Material in working memory is readily available. Otherwise considerable search and effort may be required.
Interpretation substitutes for learning. How easy it is to interpret knowledge in the world depends upon the skill of the designer.	Requires learning, which can be considerable. Learning is made easier if there is meaning or structure to the material or if there is a good conceptual model.
Slowed by the need to find and interpret the knowledge.	Can be efficient, especially if so well-learned that it is automated.
Ease of use at first encounter is high.	Ease of use at first encounter is low.
Can be ugly and inelegant, especially if there is a need to maintain a lot of knowledge. This can lead to clutter. Here is where the skills of the graphics and industrial designer play major roles.	Nothing needs to be visible, which gives more freedom to the designer. This leads to cleaner, more pleasing appearance—at the cost of ease of use at first encounter, learning, and remembering.

is ephemeral: here now, gone later. We can't count on something being present in mind at any particular time, unless it is triggered by some external event or unless we deliberately keep it in mind through constant repetition (which then prevents us from having other conscious thoughts). Out of sight, out of mind.

As we move away from many physical aids, such as printed books and magazines, paper notes, and calendars, much of what we use today as knowledge in the world will become invisible. Yes, it will all be available on display screens, but unless the screens always show this material, we will have added to the burden of memory in the head. We may not have to remember all the details of the information stored away for us, but we will have to remember that it is there, that it needs to be redisplayed at the appropriate time for use or for reminding.

Memory in Multiple Heads, Multiple Devices

If knowledge and structure in the world can combine with knowledge in the head to enhance memory performance, why not use the knowledge in multiple heads, or in multiple devices?

Most of us have experienced the power of multiple minds in remembering things. You are with a group of friends trying to remember the name of a movie, or perhaps a restaurant, and failing. But others try to help. The conversation goes something like this:

"That new place where they grill meat"

"Oh, the Korean barbecue on Fifth Street?"

"No, not Korean, South American, um,"

"Oh, yeah, Brazilian, it's what's its name?"

"Yes, that's the one!"

"Pampas something."

"Yes, Pampas Chewy. Um, Churry, um,"

"Churrascaria. Pampas Churrascaria."

How many people are involved? It could be any number, but the point is that each adds their bit of knowledge, slowly constraining the choices, recalling something that no single one of them could have done alone. Daniel Wegner, a Harvard professor of psychology, has called this "transactive memory."

Of course, we often turn to technological aids to answer our questions, reaching for our smart devices to search our electronic resources and the Internet. When we expand from seeking aids from other people to seeking aids from our technologies, which Wegner labels as "cybermind," the principle is basically the same. The cybermind doesn't always produce the answer, but it can produce sufficient clues so that we can generate the answer. Even where the technology produces the answer, it is often buried in a list of potential answers, so we have to use our own knowledge—or the knowledge of our friends—to determine which of the potential items is the correct one.

What happens when we rely too much upon external knowledge, be it knowledge in the world, knowledge of friends, or knowledge provided by our technology? On the one hand, there no such thing as "too much." The more we learn to use these resources, the better our performance. External knowledge is a powerful tool for enhanced intelligence. On the other hand, external knowledge is often erroneous: witness the difficulties of trusting online sources and the controversies that arise over Wikipedia entries. It doesn't matter where our knowledge comes from. What matters is the quality of the end result.

In an earlier book, *Things That Make Us Smart,* I argued that it is this combination of technology and people that creates superpowerful beings. Technology does not make us smarter. People do not make technology smart. It is the combination of the two, the person plus the artifact, that is smart. Together, with our tools, we are a powerful combination. On the other hand, if we are suddenly without these external devices, then we don't do very well. In many ways, we do become less smart.

Take away their calculator, and many people cannot do arithmetic. Take away a navigation system, and people can no longer get around, even in their own cities. Take away a phone's or computer's address book, and people can no longer reach their friends (in my case, I can no longer remember my own phone number).

Without a keyboard, I can't write. Without a spelling corrector, I can't spell.

What does all of this mean? Is this bad or good? It is not a new phenomenon. Take away our gas supply and electrical service and we might starve. Take away our housing and clothes and we might freeze. We rely on commercial stores, transportation, and government services to provide us with the essentials for living. Is this bad?

The partnership of technology and people makes us smarter, stronger, and better able to live in the modern world. We have become reliant on the technology and we can no longer function without it. The dependence is even stronger today than ever before, including mechanical, physical things such as housing, clothing, heating, food preparation and storage, and transportation. Now this range of dependencies is extended to information services as well: communication, news, entertainment, education, and social interaction. When things work, we are informed, comfortable, and effective. When things break, we may no longer be able to function. This dependence upon technology is very old, but every decade, the impact covers more and more activities.

Natural Mapping

Mapping, a topic from Chapter 1, provides a good example of the power of combining knowledge in the world with that in the head. Did you ever turn the wrong burner of a stove on or off? You would think that doing it correctly would be an easy task. A simple control turns the burner on, controls the temperature, and allows the burner to be turned off. In fact, the task appears to be so simple that when people do it wrong, which happens more frequently than you might have thought, they blame themselves: "How could I be so stupid as to do this simple task wrong?" they think to themselves. Well, it isn't so simple, and it is not their fault: even as simple a device as the everyday kitchen stove is frequently badly designed, in a way that guarantees the errors.

Most stoves have only four burners and four controls in one-to-one correspondence. Why is it so hard to remember four things?

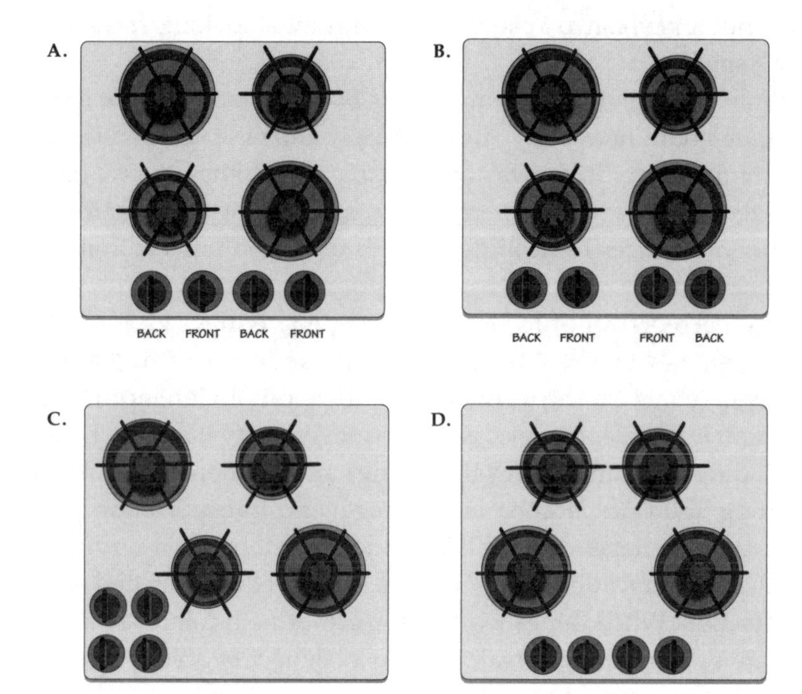

FIGURE 3.2. Mappings of Stove Controls with Burners. With the traditional arrangement of stove burners shown in Figures A and B, the burners are arranged in a rectangle and the controls in a linear line. Usually there is a partial natural mapping, with the left two controls operating the left burners and the right two controls operating the right burners. Even so, there are four possible mappings of controls to burners, all four of which are used on commercial stoves. The only way to know which control works which burner is to read the labels. But if the controls were also in a rectangle (Figure C) or the burners staggered (Figure D), no labels would be needed. Learning would be easy; errors would be reduced.

In principle, it should be easy to remember the relationship between the controls and the burners. In practice, however, it is almost impossible. Why? Because of the poor mappings between the controls and the burners. Look at Figure 3.2, which depicts four possible mappings between the four burners and controls. Figures 3.2A and B show how not to map one dimension onto two. Figures 3.2C and D show two ways of doing it properly: arrange the controls in two dimensions (C) or stagger the burners (D) so they can be ordered left to right.

To make matters worse, stove manufacturers cannot agree upon what the mapping should be. If all stoves used the same arrangement of controls, even if it is unnatural, everyone could learn it once and forever after get things right. As the legend of Figure 3.2 points out, even if the stove manufacturer is nice enough to ensure that each pair of controls operates the pair of burners on its side, there are still four possible mappings. All four are in common use. Some stoves arrange the controls in a vertical line, giving even more possible mappings. Every stove seems to be different. Even different stoves from the same manufacturer differ. No wonder people have trouble, leading their food to go uncooked, and in the worst cases, leading to fire.

Natural mappings are those where the relationship between the controls and the object to be controlled (the burners, in this case) is obvious. Depending upon circumstances, natural mappings will employ spatial cues. Here are three levels of mapping, arranged in decreasing effectiveness as memory aids:

- **Best mapping:** Controls are mounted directly on the item to be controlled.
- **Second-best mapping:** Controls are as close as possible to the object to be controlled.
- **Third-best mapping:** Controls are arranged in the same spatial configuration as the objects to be controlled.

In the ideal and second-best cases, the mappings are indeed clear and unambiguous.

Want excellent examples of natural mapping? Consider gesture-controlled faucets, soap dispensers, and hand dryers. Put your hands under the faucet or soap dispenser and the water or soap appears. Wave your hand in front of the paper towel dispenser and out pops a new towel, or in the case of blower-controlled hand dryers, simply put your hands beneath or into the dryer and the drying air turns on. Mind you, although the mappings of these devices are appropriate, they do have problems. First, they often lack signifiers, hence they lack discoverability. The controls are often invisible, so we sometimes put our hands under faucets expecting to receive water, but wait in vain: these are mechanical faucets that require handle turning. Or the water turns on and then stops, so we wave our hands up and down, hoping to find the precise location where the water turns on. When I wave my hand in front of the towel dispenser but get no towel, I do not know whether this means the dispenser is broken or out of towels; or that I did the waving wrong, or in the wrong place; or that maybe this doesn't work by gesture, but I must push, pull, or turn something. The lack of signifiers is a real drawback. These devices aren't perfect, but at least they got the mapping right.

In the case of stove controls, it is obviously not possible to put the controls directly on the burners. In most cases, it is also dangerous to put the controls adjacent to the burners, not only for fear of burning the person using the stove, but also because it would interfere with the placement of cooking utensils. Stove controls are usually situated on the side, back, or front panel of the stove, in which case they ought to be arranged in spatial harmony with the burners, as in Figures 3.2 C and D.

With a good natural mapping, the relationship of the controls to the burner is completely contained in the world; the load on human memory is much reduced. With a bad mapping, however, a burden is placed upon memory, leading to more mental effort and a higher chance of error. Without a good mapping, people new to the stove cannot readily determine which burner goes with which control and even frequent users will still occasionally err.

Why do stove designers insist on arranging the burners in a two-dimensional rectangular pattern, and the controls in a one-dimensional row? We have known for roughly a century just how bad such an arrangement is. Sometimes the stove comes with clever little diagrams to indicate which control works which burner. Sometimes there are labels. But the proper natural mapping requires no diagrams, no labels, and no instructions.

The irony about stove design is that it isn't hard to do right. Textbooks of ergonomics, human factors, psychology, and industrial engineering have been demonstrating both the problems and the solutions for over fifty years. Some stove manufacturers do use good designs. Oddly, sometimes the best and the worst designs are manufactured by the same companies and are illustrated side by side in their catalogs. Why do users still purchase stoves that cause so much trouble? Why not revolt and refuse to buy them unless the controls have an intelligent relationship to the burners?

The problem of the stovetop may seem trivial, but similar mapping problems exist in many situations, including commercial and industrial settings, where selecting the wrong button, dial, or lever can lead to major economic impact or even fatalities.

In industrial settings good mapping is of special importance, whether it is a remotely piloted airplane, a large building crane where the operator is at a distance from the objects being manipulated, or even in an automobile where the driver might wish to control temperature or windows while driving at high speeds or in crowded streets. In these cases, the best controls usually are spatial mappings of the controls to the items being controlled. We see this done properly in most automobiles where the driver can operate the windows through switches that are arranged in spatial correspondence to the windows.

Usability is not often thought about during the purchasing process. Unless you actually test a number of units in a realistic environment, doing typical tasks, you are not likely to notice the ease or difficulty of use. If you just look at something, it appears straightforward enough, and the array of wonderful features seems to be a virtue. You may not realize that you won't be able to figure out how to use those features. I urge you to test products before you buy them. Before purchasing a new stovetop, pretend you are cooking a meal. Do it right there in the store. Do not be afraid to make mistakes or ask stupid questions. Remember, any problems you have are probably the design's fault, not yours.

A major obstacle is that often the purchaser is not the user. Appliances may be in a home when people move in. In the office, the purchasing department orders equipment based upon such factors as price, relationships with the supplier, and perhaps reliability: usability is seldom considered. Finally, even when the purchaser is the end user, it is sometimes necessary to trade off one desirable feature for an undesirable one. In the case of my family's stove, we did not like the arrangement of controls, but we bought the stove anyway: we traded off the layout of the burner controls for another design feature that was more important to us and available only from one manufacturer. But why should we have to make a tradeoff? It wouldn't be hard for all stove manufacturers to use natural mappings, or at the least, to standardize their mappings.

Culture and Design: Natural Mappings Can Vary with Culture

I was in Asia, giving a talk. My computer was connected to a projector and I was given a remote controller for advancing through the illustrations for my talk. This one had two buttons, one above the other. The title was already displayed on the screen, so when I started, all I had to do was to advance to the first photograph in my presentation, but when I pushed the upper button, to my amazement I went backward through my illustrations, not forward.

"How could this happen?" I wondered. To me, top means forward; bottom, backward. The mapping is clear and obvious. If the buttons had been side by side, then the control would have been ambiguous: which comes first, right or left? This controller appeared to use an appropriate mapping of top and bottom. Why was it working backward? Was this yet another example of poor design?

I decided to ask the audience. I showed them the controller and asked: "To get to my next picture, which button should I push, the top or the bottom?" To my great surprise, the audience was split in their responses. Many thought that it should be the top button, just as I had thought. But a large number thought it should be the bottom.

What's the correct answer? I decided to ask this question to my audiences around the world. I discovered that they, too, were split in their opinions: some people firmly believe that it is the top button and some, just as firmly, believe it is the bottom button. Everyone is surprised to learn that someone else might think differently.

I was puzzled until I realized that this was a point-of-view problem, very similar to the way different cultures view time. In some cultures, time is represented mentally as if it were a road stretching out ahead of the person. As a person moves through time, the person moves forward along the time line. Other cultures use the same representation, except now it is the person who is fixed and it is time that moves: an event in the future moves toward the person.

This is precisely what was happening with the controller. Yes, the top button does cause something to move forward, but the question is, what is moving? Some people thought that the person would move through the images, other people thought the images would move. People who thought that they moved through the images wanted the top button to indicate the next one. People who thought it was the illustrations that moved would get to the next image by pushing the bottom button, causing the images to move toward them.

Some cultures represent the time line vertically: up for the future, down for the past. Other cultures have rather different views. For example, does the future lie ahead or behind? To most of us, the question makes no sense: of course, the future lies ahead—the past is behind us. We speak this way, discussing the "arrival" of the future; we are pleased that many unfortunate events of the past have been "left behind."

But why couldn't the past be in front of us and the future behind? Does that sound strange? Why? We can see what is in front of us, but not what is behind, just as we can remember what happened in the past, but we can't remember the future. Not only that, but we can remember recent events much more clearly than long-past events, captured neatly by the visual metaphor in which the past lines up before us, the most recent events being the closest so that they are clearly perceived (remembered), with long-past events far in the distance, remembered and perceived with difficulty. Still sound weird? This is how the South American Indian group, the Aymara, represent time. When they speak of the future, they use the phrase *back days* and often gesture behind them. Think about it: it is a perfectly logical way to view the world.

If time is displayed along a horizontal line, does it go from left to right or right to left? Either answer is correct because the choice is arbitrary, just as the choice of whether text should be strung along the page from left to right or right to left is arbitrary. The choice of text direction also corresponds to people's preference for time direction. People whose native language is Arabic or Hebrew prefer time to flow from right to left (the future being toward the left), whereas those who use a left-to-right writing system have time flowing in the same direction, so the future is to the right.

But wait: I'm not finished. Is the time line relative to the person or relative to the environment? In some Australian Aborigine societies, time moves relative to the environment based on the direction in which the sun rises and sets. Give people from this community a set of photographs structured in time (for example, photographs of a person at different ages or a child eating some food) and ask them to order the photographs in time. People from technological cultures would order the pictures from left to right, most recent photo to the right or left, depending upon how their printed language was written. But people from these Australian communities would order them east to west, most recent to the west. If the person were facing south, the photo would be ordered left to right. If the person were facing north, the photos would be ordered right to left. If the person were facing west, the photos would be ordered along a vertical line extending from the body outward, outwards being the most recent. And, of course, were the person facing east, the photos would also be on a line extending out from the body, but with the most recent photo closest to the body.

The choice of metaphor dictates the proper design for interaction. Similar issues show up in other domains. Consider the standard problem of scrolling the text in a computer display. Should the scrolling control move the text or the window? This was a fierce debate in the early years of display terminals, long before the development of modern computer systems. Eventually, there was mutual agreement that the cursor arrow keys—and then, later on, the mouse—would follow the moving window metaphor. Move the window down to see more text at the bottom of the screen. What this meant in practice is that to see more text at the bottom of the screen, move the mouse down, which moves the window down, so that the text moves up: the mouse and the text move in opposite directions. With the moving text metaphor, the mouse and the text move in the same directions: move the mouse up and the text moves up. For over two decades, everyone moved the scrollbars and mouse down in order to make the text move up.

But then smart displays with touch-operated screens arrived. Now it was only natural to touch the text with the fingers and move it up, down, right, or left directly: the text moved in the same direction as the fingers. The moving text metaphor became prevalent. In fact, it was no longer thought of as a metaphor: it was real. But as people switched back and forth between traditional computer systems that used the moving window metaphor and touch-screen systems that used the moving text model, confusion reigned. As a result, one major manufacturer of both computers and smart screens, Apple, switched everything to the moving text model, but no other company followed Apple's lead. As I write this, the confusion still exists. How will it end? I predict the demise of the moving window metaphor: touch-screens and control pads will dominate, which will cause the moving text model to take over. All systems will move the hands or controls in the same direction as they wish the screen images to move. Predicting technology is relatively easy compared to predictions of human behavior, or in this case, the adoption of societal conventions. Will this prediction be true? You will be able to judge for yourself.

Similar issues occurred in aviation with the pilot's attitude indicator, the display that indicates the airplane's orientation (roll or bank and pitch). The instrument shows a horizontal line to indicate the horizon with a silhouette of an airplane seen from behind. If the wings are level and on a line with the horizon, the airplane is flying in level flight. Suppose the airplane turns to the left, so it banks (tilts) left. What should the display look like? Should it show a left-tilting airplane against a fixed horizon, or a fixed airplane against a right-tilting horizon? The first is correct from the viewpoint of someone watching the airplane from behind, where the horizon is always horizontal: this type of display is called *outside-in*. The second is correct from the viewpoint of the pilot, where the airplane is always stable and fixed in position, so that when the airplane banks, the horizon tilts: this type of display is called *inside-out*.

In all these cases, every point of view is correct. It all depends upon what you consider to be moving. What does all this mean for design? What is natural depends upon point of view, the choice of metaphor, and therefore, the culture. The design difficulties occur when there is a switch in metaphors. Airplane pilots have to undergo training and testing before they are allowed to switch from one set of instruments (those with an outside-in metaphor, for example) to the other (those with the inside-out metaphor). When countries decided to switch which side of the road cars would drive on, the temporary confusion that resulted was dangerous. (Most places that switched moved from left-side driving to right-side, but a few, notably Okinawa, Samoa, and East Timor, switched from right to left.) In all these cases of convention switches, people eventually adjusted. It is possible to break convention and switch metaphors, but expect a period of confusion until people adapt to the new system.

CHAPTER FOUR

KNOWING WHAT TO DO: CONSTRAINTS, DISCOVERABILITY, AND FEEDBACK

How do we determine how to operate something that we have never seen before? We have no choice but to combine knowledge in the world with that in the head. Knowledge in the world includes perceived affordances and signifiers, the mappings between the parts that appear to be controls or places to manipulate and the resulting actions, and the physical constraints that limit what can be done. Knowledge in the head includes conceptual models; cultural, semantic, and logical constraints on behavior; and analogies between the current situation and previous experiences with other situations. Chapter 3 was devoted to a discussion of how we acquire knowledge and use it. There, the major emphasis was upon the knowledge in the head. This chapter focuses upon the knowledge in the world: how designers can provide the critical information that allows people to know what to do, even when experiencing an unfamiliar device or situation.

Let me illustrate with an example: building a motorcycle from a Lego set (a children's construction toy). The Lego motorcycle shown in Figure 4.1 has fifteen pieces, some rather specialized. Of those fifteen pieces, only two pairs are alike—two rectangles with the word *police* on them, and the two hands of

FIGURE 4.1. Lego Motorcycle. The toy Lego motorcycle is shown assembled (A) and in pieces (B). It has fifteen pieces so cleverly constructed that even an adult can put them together. The design exploits constraints to specify just which pieces fit where. Physical constraints limit alternative placements. Cultural and semantic constraints provide the necessary clues for further decisions. For example, cultural constraints dictate the placement of the three lights (red, blue, and yellow) and semantic constraints stop the user from putting the head backward on the body or the pieces labeled "police" upside down.

the policeman. Other pieces match one another in size and shape but are different colors. So, a number of the pieces are physically interchangeable—that is, the physical constraints are not sufficient to identify where they go—but the appropriate role for every single piece of the motorcycle is still unambiguously determined. How? By combining cultural, semantic, and logical constraints with the physical ones. As a result, it is possible to construct the motorcycle without any instructions or assistance.

In fact, I did the experiment. I asked people to put together the parts; they had never seen the finished structure and were not even told that it was a motorcycle (although it didn't take them long to figure this out). Nobody had any difficulty.

The visible affordances of the pieces were important in determining just how they fit together. The cylinders and holes characteristic of Lego suggested the major construction rule. The sizes and shapes of the parts suggested their operation. Physical constraints limited what parts would fit together. Cultural and semantic constraints provided strong restrictions on what would make sense for all but one of the remaining pieces, and with just one piece left and only one place it could possibly go, simple logic dictated the placement. These four classes of constraints—physical, cultural, semantic, and logical—seem to be universal, appearing in a wide variety of situations.

Constraints are powerful clues, limiting the set of possible actions. The thoughtful use of constraints in design lets people readily determine the proper course of action, even in a novel situation.

Four Kinds of Constraints: Physical, Cultural, Semantic, and Logical

PHYSICAL CONSTRAINTS

Physical limitations constrain possible operations. Thus, a large peg cannot fit into a small hole. With the Lego motorcycle, the windshield would fit in only one place. The value of physical constraints is that they rely upon properties of the physical world for their operation; no special training is necessary. With the proper use of physical constraints, there should be only a limited number of possible actions—or, at least, desired actions can be made obvious, usually by being especially salient.

Physical constraints are made more effective and useful if they are easy to see and interpret, for then the set of actions is restricted before anything has been done. Otherwise, a physical constraint prevents a wrong action from succeeding only after it has been tried.

The traditional cylindrical battery, Figure 4.2A, lacks sufficient physical constraints. It can be put into battery compartments in two orientations: one that is correct, the other of which can damage the equipment. The instructions in Figure 4.2B show that polarity is important, yet the inferior signifiers inside the battery compartment makes it very difficult to determine the proper orientation for the batteries.

Why not design a battery with which it would be impossible to make an error: use physical constraints so that the battery will fit only if properly oriented. Alternatively, design the battery or the electrical contacts so that orientation doesn't matter.

Figure 4.3 shows a battery that has been designed so that orientation is irrelevant. Both ends of the battery are identical, with the

A.

B.

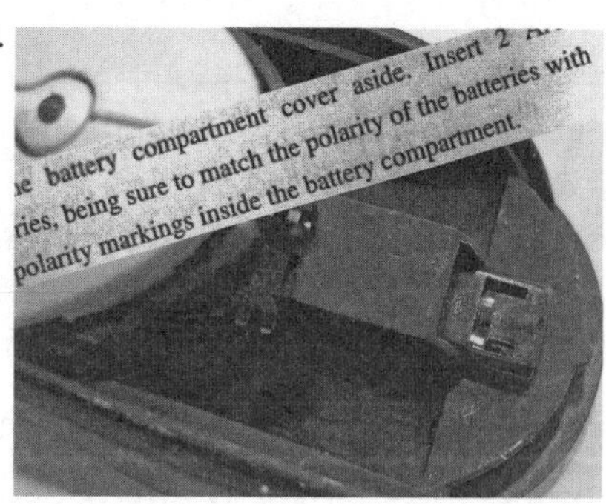

FIGURE 4.2. Cylindrical Battery: Where Constraints Are Needed. Figure A shows the traditional cylindrical battery that requires correct orientation in the slot to work properly (and to avoid damaging the equipment). But look at Figure B, which shows where two batteries are to be installed. The instructions from the manual are shown as an overlay to the photograph. They seem simple, but can you see into the dark recess to figure out which end of each battery goes where? Nope. The lettering is black against black: slightly raised shapes in the dark plastic.

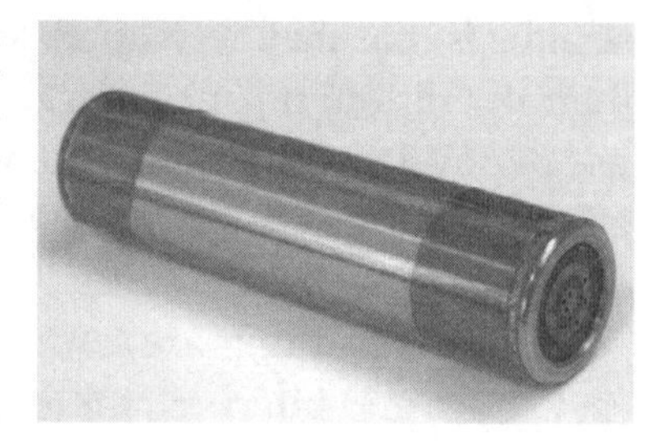

FIGURE 4.3. Making Battery Orientation Irrelevant. This photograph shows a battery whose orientation doesn't matter; it can be inserted into the equipment in either possible direction. How? Each end of the battery has the same three concentric rings, with the center one on both ends being the "plus" terminal and the middle one being the "minus" terminal.

positive and negative terminals for the battery being its center and middle rings, respectively. The contact for the positive polarity is designed so it contacts only the center ring. Similarly, the contact for negative polarity touches only the middle ring. Although this seems to solve the problem, I have only seen this one example of such a battery: they are not widely available or used.

Another alternative is to invent battery contacts that allow our existing cylindrical batteries to be inserted in either orientation yet still work properly: Microsoft has invented this kind of contact, which it calls InstaLoad, and is attempting to convince equipment manufacturers to use it.

A third alternative is to design the shape of the battery so that it can fit in only one way. Most plug-in components do this well, using shapes, notches, and protrusions to constrain insertion to a single orientation. So why can't our everyday batteries be the same?

Why does inelegant design persist for so long? This is called the *legacy problem,* and it will come up several times in this book. Too many devices use the existing standard—that is the legacy. If the symmetrical cylindrical battery were changed, there would also have to be a major change in a huge number of products. The new batteries would not work in older equipment, nor the old batteries in new equipment. Microsoft's design of contacts would allow us to continue to use the same batteries we are used to, but the products would have to switch to the new contacts. Two years after Microsoft's introduction of InstaLoad, despite positive press, I could find no products that use them—not even Microsoft products.

Locks and keys suffer from a similar problem. Although it is usually easy to distinguish the smooth top part of a key from its jagged underside, it is difficult to tell from the lock just which orientation of the key is required, especially in dark environments. Many electrical and electronic plugs and sockets have the same problem. Although they do have physical constraints to prevent improper insertion, it is often extremely difficult to perceive their correct orientation, especially when keyholes and electronic sockets are in difficult-to-reach, dimly lit locations. Some devices, such as USB plugs, are constrained, but the constraint is so subtle that it takes much fussing and fumbling to find the correct orientation. Why aren't all these devices orientation insensitive?

It is not difficult to design keys and plugs that work regardless of how they are inserted. Automobile keys that are insensitive to the orientation have long existed, but not all manufacturers use them. Similarly, many electrical connectors are insensitive to orientation, but again, only a few manufacturers use them. Why the resistance? Some of it results from the legacy concerns about the expense of massive change. But much seems to be a classic example of corporate thinking: "This is the way we have always done things. We don't care about the customer." It is, of course, true that difficulty in inserting keys, batteries, or plugs is not a big enough issue to affect the decision of whether to purchase something, but still, the lack of attention to customer needs on even simple things is often symptomatic of larger issues that have greater impact.

Note that a superior solution would be to solve the fundamental need—solving the root need. After all, we don't really care about keys and locks: what we need is some way of ensuring that only authorized people can get access to whatever is being locked. Instead of redoing the shapes of physical keys, make them irrelevant. Once this is recognized, a whole set of solutions present themselves: combination locks that do not require keys, or keyless locks that can be operated only by authorized people. One method is through possession of an electronic wireless device, such as the identification badges that unlock doors when they are moved close to a sensor, or automobile keys that can stay in the pocket or carrying case. Biometric devices could identify the person through face or voice recognition, fingerprints, or other biometric measures, such as iris patterns. This approach is discussed in Chapter 3, page 91.

CULTURAL CONSTRAINTS

Each culture has a set of allowable actions for social situations. Thus, in our own culture we know how to behave in a restaurant—even one we have never been to before. This is how we manage to cope when our host leaves us alone in a strange room, at a strange party, with strange people. And this is why we sometimes feel frustrated, so incapable of action, when we are confronted with a restaurant or group of people from an unfamiliar culture, where our normally accepted behavior is clearly inappropriate and frowned upon. Cultural issues are at the root of many of the problems we have with new machines: there are as yet no universally accepted conventions or customs for dealing with them.

Those of us who study these things believe that guidelines for cultural behavior are represented in the mind by schemas, knowledge structures that contain the general rules and information necessary for interpreting situations and for guiding behavior. In some stereotypical situations (for example, in a restaurant), the schemas may be very specialized. Cognitive scientists Roger Schank and

Bob Abelson proposed that in these cases we follow "scripts" that can guide the sequence of behavior. The sociologist Erving Goffman calls the social constraints on acceptable behavior "frames," and he shows how they govern behavior even when a person is in a novel situation or novel culture. Danger awaits those who deliberately violate the frames of a culture.

The next time you are in an elevator, try violating cultural norms and see how uncomfortable that makes you and the other people in the elevator. It doesn't take much: Stand facing the rear. Or look directly at some of the passengers. In a bus or streetcar, give your seat to the next athletic-looking person you see (the act is especially effective if you are elderly, pregnant, or disabled).

In the case of the Lego motorcycle of Figure 4.1, cultural constraints determine the locations of the three lights of the motorcycle, which are otherwise physically interchangeable. Red is the culturally defined standard for a brake light, which is placed in the rear. And a police vehicle often has a blue flashing light on top. As for the yellow piece, this is an interesting example of cultural change: few people today remember that yellow used to be a standard headlight color in Europe and a few other locations (Lego comes from Denmark). Today, European and North American standards require white headlights. As a result, figuring out that the yellow piece represents a headlight on the front of the motorcycle is no longer as easy as it used to be. Cultural constraints are likely to change with time.

SEMANTIC CONSTRAINTS

Semantics is the study of meaning. Semantic constraints are those that rely upon the meaning of the situation to control the set of possible actions. In the case of the motorcycle, there is only one meaningful location for the rider, who must sit facing forward. The purpose of the windshield is to protect the rider's face, so it must be in front of the rider. Semantic constraints rely upon our knowledge of the situation and of the world. Such knowledge can be a powerful and important clue. But just as cultural constraints can change with time, so, too, can semantic ones. Extreme sports push the boundaries of what we think of as meaningful and sensible. New technologies change the meanings of things. And creative people continually change how we interact with our technologies and one another. When cars become fully automated, communicating among themselves with wireless networks, what will be the meaning of the red lights on the rear of the auto? That the car is braking? But for whom would the signal be intended? The other cars would already know. The red light would become meaningless, so it could either be removed or it could be redefined to indicate some other condition. The meanings of today may not be the meanings of the future.

LOGICAL CONSTRAINTS

The blue light of the Lego motorcycle presents a special problem. Many people had no knowledge that would help, but after all the other pieces had been placed on the motorcycle, there was only one piece left, only one possible place to go. The blue light was logically constrained.

Logical constraints are often used by home dwellers who undertake repair jobs. Suppose you take apart a leaking faucet to replace a washer, but when you put the faucet together again, you discover a part left over. Oops, obviously there was an error: the part should have been installed. This is an example of a logical constraint.

The natural mappings discussed in Chapter 3 work by providing logical constraints. There are no physical or cultural principles here; rather, there is a logical relationship between the spatial or functional layout of components and the things that they affect or are affected by. If two switches control two lights, the left switch should work the left light; the right switch, the right light. If the orientation of the lights and the switches differ, the natural mapping is destroyed.

CULTURAL NORMS, CONVENTIONS, AND STANDARDS

Every culture has its own conventions. Do you kiss or shake hands when meeting someone? If kissing, on which cheek, and how many times? Is it an air kiss or an actual one? Or perhaps you bow, junior person first, and lowest. Or raise hands, or perhaps press them together. Sniff? It is possible to spend a fascinating hour on the Internet exploring the different forms of greetings used by different cultures. It is also amusing to watch the consternation when people from more cool, formal countries first encounter people from warm-hearted, earthy countries, as one tries to bow and shake hands and the other tries to hug and kiss even total strangers. It is not so amusing to be one of those people: being hugged or kissed while trying to shake hands or bow. Or the other way around. Try kissing someone's cheek three times (left, right, left) when the person expects only one. Or worse, where he or she expects a handshake. Violation of cultural conventions can completely disrupt an interaction.

Conventions are actually a form of cultural constraint, usually associated with how people behave. Some conventions determine what activities should be done; others prohibit or discourage actions. But in all cases, they provide those knowledgeable of the culture with powerful constraints on behavior.

Sometimes these conventions are codified into international standards, sometimes into laws, and sometimes both. In the early days of heavily traveled streets, whether by horses and buggies or by automobiles, congestion and accidents arose. Over time, conventions developed about which side of the road to drive on, with different conventions in different countries. Who had precedence at crossings? The first person to get there? The vehicle or person on the right, or the person with the highest social status? All of these conventions have applied at one time or another. Today, worldwide standards govern many traffic situations: Drive on only one side of the street. The first car into an intersection has precedence. If both arrive at the same time, the car on the right (or left) has precedence. When merging traffic lanes, alternate cars—one from that lane, then one from this. The last rule is more of an informal convention: it is not part of any rule book that I am aware of, and although it is very nicely obeyed in the California streets on which I drive, the very concept would seem strange in some parts of the world.

Sometimes conventions clash. In Mexico, when two cars approach a narrow, one-lane bridge from opposite directions, if a car blinks its headlights, it means, "I got here first and I'm going over the bridge." In England, if a car blinks its lights, it means, "I see you: please go first." Either signal is equally appropriate and useful, but not if the two drivers follow different conventions. Imagine a Mexican driver meeting an English driver in some third country. (Note that driving experts warn against using headlight blinks as signals because even within any single country, either interpretation is held by many drivers, none of whom imagines someone else might have the opposite interpretation.)

Ever get embarrassed at a formal dinner party where there appear to be dozens of utensils at each place setting? What do you do? Do you drink that nice bowl of water or is it for dipping your fingers to clean them? Do you eat a chicken drumstick or slice of pizza with your fingers or with a knife and fork?

Do these issues matter? Yes, they do. Violate conventions and you are marked as an outsider. A rude outsider, at that.

Applying Affordances, Signifiers, and Constraints to Everyday Objects

Affordances, signifiers, mappings, and constraints can simplify our encounters with everyday objects. Failure to properly deploy these cues leads to problems.

THE PROBLEM WITH DOORS

In Chapter 1 we encountered the sad story of my friend who was trapped between sets of glass doors at a post office, trapped because there were no clues to the doors' operation. To operate a door, we have to find the side that opens and the part to be manipulated; in other words, we need to figure out what to do and where to do it. We expect to find some visible signal, a signifier, for the correct operation: a plate, an extension, a hollow, an indentation—something that allows the hand to touch, grasp, turn, or fit into. This tells us where to act. The next step is to figure out how: we must determine what operations are permitted, in part by using the signifiers, in part guided by constraints.

Doors come in amazing variety. Some open only if a button is pushed, and some don't indicate how to open at all, having neither buttons, nor hardware, nor any other sign of their operation. The door might be operated with a foot pedal. Or maybe it is voice operated, and we must speak the magic phrase ("Open Simsim!"). In addition, some doors have signs on them, to pull, push, slide, lift, ring a bell, insert a card, type a password, smile, rotate, bow, dance, or, perhaps, just ask. Somehow, when a device as simple as a door has to have a sign telling you whether to pull, push, or slide, then it is a failure, poorly designed.

Consider the hardware for an unlocked door. It need not have any moving parts: it can be a fixed knob, plate, handle, or groove. Not only will the proper hardware operate the door smoothly, but it will also indicate just how the door is to be operated: it will incorporate clear and unambiguous clues—signifiers. Suppose the door opens by being pushed. The easiest way to indicate this is to have a plate at the spot where the pushing should be done.

Flat plates or bars can clearly and unambiguously signify both the proper action and its location, for their affordances constrain the possible actions to that of pushing. Remember the discussion of the fire door and its panic bar in Chapter 2 (Figure 2.5, page 60)? The panic bar, with its large horizontal surface, often with a secondary color on the part intended to be pushed, provides a good example of an unambiguous signifier. It very nicely constrains improper behavior when panicked people press against the door as they attempt to flee a fire. The best push bars offer both visible affordances that act as physical constraints on the action, and also a visible signifier, thereby unobtrusively specifying *what* to do and *where* to do it.

Some doors have appropriate hardware, well placed. The outside door handles of most modern automobiles are excellent examples of design. The handles are often recessed receptacles that simultaneously indicate the place and mode of action. Horizontal slits guide the hand into a pulling position; vertical slits signal a sliding motion. Strangely enough, the inside door handles for automobiles tell a different story. Here, the designer has faced a different kind of problem, and the appropriate solution has not yet been found. As a result, although the outside door handles of cars are often excellent, the inside ones are often difficult to find, hard to figure out how to operate, and awkward to use.

From my experience, the worst offenders are cabinet doors. It is sometimes not even possible to determine where the doors are, let alone whether and how they are slid, lifted, pushed, or pulled. The focus on aesthetics may blind the designer (and the purchaser) to the lack of usability. A particularly frustrating design is that of the cabinet door that opens outward by being pushed inward. The push releases the catch and energizes a spring, so that when the hand is taken away, the door springs open. It's a very clever design, but most puzzling to the first-time user. A plate would be the appropriate signal, but designers do not wish to mar the smooth surface of the door. One of the cabinets in my home has one of these latches in its glass door. Because the glass affords visibility of the shelves inside, it is obvious that there is no room for the door to open inward; therefore, to push the door seems contradictory. New and infrequent users of this door usually reject pushing and open it by pulling, which often requires them to use fingernails, knife blades, or more ingenious methods to pry it open. A similar, counterintuitive type of design was the source of my difficulties in emptying the dirty water from my sink in a London hotel (Figure 1.4, page 17).

Appearances deceive. I have seen people trip and fall when they attempted to push open a door that worked automatically, the door opening inward just as they attempted to push against it. On most subway trains, the doors open automatically at each station. Not so in Paris. I watched someone on the Paris Métro try to get off the train and fail. When the train came to his station, he got up and stood patiently in front of the door, waiting for it to open. It never opened. The train simply started up again and went on to the next station. In the Métro, you have to open the doors yourself by pushing a button, or depressing a lever, or sliding them (depending upon which kind of car you happen to be on). In some transit systems, the passenger is supposed to operate the door, but in others this is forbidden. The frequent traveler is continually confronted with this kind of situation: the behavior that is appropriate in one place is inappropriate in another, even in situations that appear to be identical. Known cultural norms can create comfort and harmony. Unknown norms can lead to discomfort and confusion.

THE PROBLEM WITH SWITCHES

When I give talks, quite often my first demonstration needs no preparation. I can count on the light switches of the room or auditorium to be unmanageable. "Lights, please," someone will say. Then fumble, fumble, fumble. Who knows where the switches are and which lights they control? The lights seem to work smoothly only when a technician is hired to sit in a control room somewhere, turning them on and off.

The switch problems in an auditorium are annoying, but similar problems in industry could be dangerous. In many control rooms, row upon row of identical-looking switches confront the operators. How do they avoid the occasional error, confusion, or accidental bumping against the wrong control? Or mis-aim? They don't. Fortunately, industrial settings are usually pretty robust. A few errors every now and then are not important—usually.

One type of popular small airplane has identical-looking switches for flaps and for landing gear, right next to one another. You might be surprised to learn how many pilots, while on the ground, have decided to raise the flaps and instead raised the wheels. This very expensive error happened frequently enough that the National Transportation Safety Board wrote a report about it. The analysts politely pointed out that the proper design principles to avoid these errors had been known for fifty years. Why were these design errors still being made?

Basic switches and controls should be relatively simple to design well. But there are two fundamental difficulties. The first is to determine what type of device they control; for example, flaps or landing gear. The second is the mapping problem, discussed extensively in Chapters 1 and 3; for example, when there are many lights and an array of switches, which switch controls which light?

The switch problem becomes serious only where there are many of them. It isn't a problem in situations with one switch, and it is only a minor problem where there are two switches. But the difficulties mount rapidly with more than two switches at the same location. Multiple switches are more likely to appear in offices, auditoriums, and industrial locations than in homes.

With complex installations, where there are numerous lights and switches, the light controls seldom fit the needs of the situation. When I give talks, I need a way to dim the light hitting the projection screen so that images are visible, but keep enough light on the audience so that they can take notes (and I can monitor their reaction to the talk). This kind of control is seldom provided. Electricians are not trained to do task analyses.

Whose fault is this? Probably nobody's. Blaming a person is seldom appropriate or useful, a point I return to in Chapter 5. The problem is probably due to the difficulties of coordinating the different professions involved in installing light controls.

FIGURE 4.4. Incomprehensible Light Switches. Banks of switches like this are not uncommon in homes. There is no obvious mapping between the switches and the lights being controlled. I once had a similar panel in my home, although with only six switches. Even after years of living in the house, I could never remember which to use, so I simply put all the switches either up (on) or down (off). How did I solve the problem? See Figure 4.5.

I once lived in a wonderful house on the cliffs of Del Mar, California, designed for us by two young, award-winning architects. The house was wonderful, and the architects proved their worth by the spectacular placement of the house and the broad windows that overlooked the ocean. But they liked spare, neat, modern design to a fault. Inside the house were, among other things, neat rows of light switches: A horizontal row of four identical switches in the front hall, a vertical column of six identical switches in the living room. "You will get used to it," the architects assured us when we complained. We never did. Figure 4.4 shows an eight-switch bank that I found in a home I was visiting. Who could remember what each does? My home only had six switches, and that was bad enough. (Photographs of the switch plate from my Del Mar home are no longer available.)

The lack of clear communication among the people and organizations constructing parts of a system is perhaps the most common cause of complicated, confusing designs. A usable design starts with careful observations of how the tasks being supported are actually performed, followed by a design process that results in a good fit to the actual ways the tasks get performed. The technical name for this method is *task analysis.* The name for the entire process is *human-centered design* (HCD), discussed in Chapter 6.

The solutions to the problem posed by my Del Mar home require the natural mappings described in Chapter 3. With six light switches mounted in a one-dimensional array, vertically on the wall, there is no way they can map naturally to the two-dimensional, horizontal placement of the lights in the ceiling. Why place the switches flat against the wall? Why not redo things? Why not place the switches horizontally, in exact analogy to the things being controlled, with a two-dimensional layout so that the switches can be placed on a floor plan of the building in exact correspondence to the areas that they control? Match the layout of the lights with the layout of the switches: the principle of natural mapping. You can see the result in Figure 4.5. We mounted a floor plan of the living room on a plate and oriented it to match the room. Switches were placed on the floor plan so that each switch was located in the area controlled

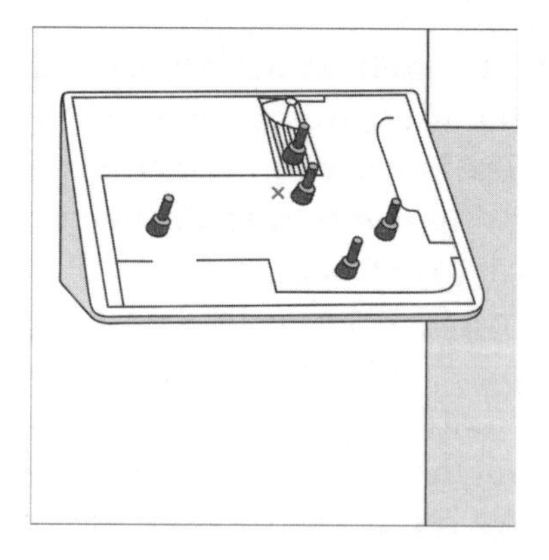

FIGURE 4.5. A Natural Mapping of Light Switches to Lights. This is how I mapped five switches to the lights in my living room. I placed small toggle switches that fit onto a plan of the home's living room, balcony, and hall, with each switch placed where the light was located. The X by the center switch indicates where this panel was located. The surface was tilted to make it easier to relate it to the horizontal arrangement of the lights, and the slope provided a natural anti-affordance, preventing people from putting coffee cups and drink containers on the controls.

by that switch. The plate was mounted with a slight tilt from the horizontal to make it easy to see and to make the mapping clear: had the plate been vertical, the mapping would still be ambiguous. The plate was tilted rather than horizontal to discourage people (us or visitors) from placing objects, such as cups, on the plate: an example of an anti-affordance. (We further simplified operations by moving the sixth switch to a different location where its meaning was clear and it did not confuse, because it stood alone.)

It is unnecessarily difficult to implement this spatial mapping of switches to lights: the required parts are not available. I had to hire a skilled technician to construct the wall-mounted box and install the special switches and control equipment. Builders and electricians need standardized components. Today, the switch boxes that are available to electricians are organized as rectangular boxes meant to hold a long, linear string of switches and to be mounted horizontally or vertically on the wall. To produce the appropriate spatial array, we would need a two-dimensional structure that could be mounted parallel to the floor, where the switches would be mounted on the top of the box, on the horizontal surface. The switch box should have a matrix of supports so that there can be free, relatively unrestricted placement of the switches in whatever pattern best suits the room. Ideally the box would use small switches, perhaps low-voltage switches that would control a separately mounted control structure that takes care of the lights (which is what I did in my home). Switches and lights could communicate wirelessly instead of through the traditional home wiring cables. Instead of the standardized light plates for today's large, bulky switches, the plates should be designed for small holes appropriate to the small switches, combined with a way of inserting a floor plan on to the switch cover.

My suggestion requires that the switch box stick out from the wall, whereas today's boxes are mounted so that the switches are flush with the wall. But these new switch boxes wouldn't have to stick out. They could be placed in indented openings in the walls: just as there is room inside the wall for the existing switch boxes, there is also room for an indented horizontal surface. Or the switches could be mounted on a little pedestal.

As a side note, in the decades that have passed since the first edition of this book was published, the section on natural mappings and the difficulties with light switches has received a very popular reception. Nonetheless, there are no commercial tools available to make it easy to implement these ideas in the home. I once tried to convince the CEO of the company whose smart home devices I had used to implement the controls of Figure 4.5, to use the idea. "Why not manufacture the components to make it easy for people to do this," I suggested. I failed.

Someday, we will get rid of the hard-wired switches, which require excessive runs of electrical cable, add to the cost and difficulties of home construction, and make remodeling of electrical circuits extremely difficult and time consuming. Instead, we will use Internet or wireless signals to connect switches to the devices to be controlled. In this way, controls could be located anywhere. They could be reconfigured or moved. We could have multiple controls for the same item, some in our phones or other portable devices. I can control my home thermostat from anywhere in the world: why can't I do the same with my lights? Some of the necessary technology does exist today in specialty shops and custom builders, but they will not come into widespread usage until major manufacturers make the necessary components and traditional electricians become comfortable with installing them. The tools for creating switch configurations that use good mapping principles could become standard and easy to apply. It will happen, but it may take considerable time.

Alas, like many things that change, new technologies will bring virtues and deficits. The controls are apt to be through touch-sensitive screens, allowing excellent natural mapping to the spatial layouts involved, but lacking the physical affordances of physical switches. They can't be operated with the side of the arm or the elbow while trying to enter a room, hands loaded with packages or cups of coffee. Touch screens are fine if the hands are free. Perhaps cameras that recognize gestures will do the job.

ACTIVITY-CENTERED CONTROLS

Spatial mapping of switches is not always appropriate. In many cases it is better to have switches that control activities: activity-centered control. Many auditoriums in schools and companies have computer-based controls, with switches labeled with such phrases as "video," "computer," "full lights," and "lecture." When carefully designed, with a good, detailed analysis of the activities to be supported, the mapping of controls to activities works extremely well: video requires a dark auditorium plus control of sound level and controls to start, pause, and stop the presentation. Projected images require a dark screen area with enough light in the auditorium so people can take notes. Lectures require some stage lights so the speaker can be seen. Activity-based controls are excellent in theory, but the practice is difficult to get right. When it is done badly, it creates difficulties.

A related but wrong approach is to be device-centered rather than activity-centered. When they are device-centered, different control screens cover lights, sound, computer, and video projection. This requires the lecturer to go to one screen to adjust the light, a different screen to adjust sound levels, and yet a different screen to advance or control the images. It is a horrible cognitive interruption to the flow of the talk to go back and forth among the screens, perhaps to pause the video in order to make a comment or answer a question. Activity-centered controls anticipate this need and put light, sound level, and projection controls all in one location.

I once used an activity-centered control, setting it to present my photographs to the audience. All worked well until I was asked a question. I paused to answer it, but wanted to raise the room lights so I could see the audience. No, the activity of giving a talk with visually presented images meant that room lights were fixed at a dim setting. When I tried to increase the light intensity, this took me out of "giving a talk" activity, so I did get the light to where I wanted it, but the projection screen also went up into the ceiling and the projector was turned off. The difficulty with activity-based controllers is handling the exceptional cases, the ones not thought about during design.

Activity-centered controls are the proper way to go, if the activities are carefully selected to match actual requirements. But even in these cases, manual controls will still be required because there will always be some new, unexpected demand that requires idiosyncratic settings. As my example demonstrates, invoking the manual settings should not cause the current activity to be canceled.

Constraints That Force the Desired Behavior

FORCING FUNCTIONS

Forcing functions are a form of physical constraint: situations in which the actions are constrained so that failure at one stage prevents the next step from happening. Starting a car has a forcing function associated with it—the driver must have some physical object that signifies permission to use the car. In the past, it was a physical key to unlock the car doors and also to be placed into the ignition switch, which allowed the key to turn on the electrical system and, if rotated to its extreme position, to activate the engine.

Today's cars have many means of verifying permission. Some still require a key, but it can stay in one's pocket or carrying case. More and more, the key is not required and is replaced by a card, phone, or some physical token that can communicate with the car. As long as only authorized people have the card (which is, of course, the same for keys), everything works fine. Electric or hybrid vehicles do not need to start the engines prior to moving the car, but the procedures are still similar: drivers must authenticate themselves by having a physical item in their possession. Because the vehicle won't start without the authentication proved by possession of the key, it is a forcing function.

Forcing functions are the extreme case of strong constraints that can prevent inappropriate behavior. Not every situation allows such strong constraints to operate, but the general principle can be extended to a wide variety of situations. In the field of safety engineering, forcing functions show up under other names, in particular as specialized methods for the prevention of accidents. Three such methods are interlocks, lock-ins, and lockouts.

INTERLOCKS

An interlock forces operations to take place in proper sequence. Microwave ovens and devices with interior exposure to high voltage use interlocks as forcing functions to prevent people from opening the door of the oven or disassembling the devices without first turning off the electric power: the interlock disconnects the power the instant the door is opened or the back is removed. In automobiles with automatic transmissions, an interlock prevents the transmission from leaving the Park position unless the car's brake pedal is depressed.

Another form of interlock is the "dead man's switch" in numerous safety settings, especially for the operators of trains, lawn mowers, chainsaws, and many recreational vehicles. In Britain, these are called the "driver's safety device." Many require that the operator hold down a spring-loaded switch to enable operation of the equipment, so that if the operator dies (or loses control), the switch will be released, stopping the equipment. Because some operators bypassed the feature by tying down the control (or placing a heavy weight on foot-operated ones), various schemes have been developed to determine that the person is really alive and alert. Some require a midlevel of pressure; some, repeated depressions and releases. Some require responses to queries. But in all cases,

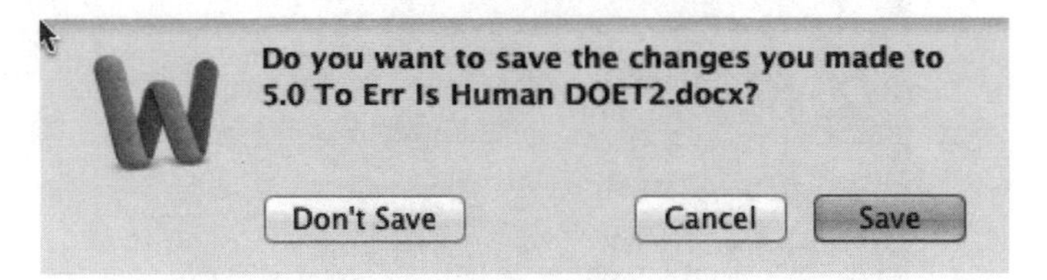

FIGURE 4.6 A Lock-In Forcing Function. This lock-in makes it difficult to exit a program without either saving the work or consciously saying not to. Notice that it is politely configured so that the desired operation can be taken right from the message.

they are examples of safety-related interlocks to prevent operation when the operator is incapacitated.

LOCK-INS

A lock-in keeps an operation active, preventing someone from prematurely stopping it. Standard lock-ins exist on many computer applications, where any attempt to exit the application without saving work is prevented by a message prompt asking whether that is what is really wanted (Figure 4. 6). These are so effective that I use them deliberately as my standard way of exiting. Rather than saving a file and then exiting the program, I simply exit, knowing that I will be given a simple way to save my work. What was once created as an error message has become an efficient shortcut.

Lock-ins can be quite literal, as in jail cells or playpens for babies, preventing a person from leaving the area.

Some companies try to lock in customers by making all their products work harmoniously with one another but be incompatible with the products of their competition. Thus music, videos, or electronic books purchased from one company may be played or read on music and video players and e-book readers made by that company, but will fail with similar devices from other manufacturers. The goal is to use design as a business strategy: the consistency within a given manufacturer means once people learn the system, they will stay with it and hesitate to change. The confusion when using a different company's system further prevents customers from

FIGURE 4.7. A Lockout Forcing Function for Fire Exit. The gate, placed at the ground floor of stairways, prevents people who might be rushing down the stairs to escape a fire from continuing into the basement areas, where they might get trapped.

changing systems. In the end, the people who must use multiple systems lose. Actually, everyone loses, except for the one manufacturer whose products dominate.

LOCKOUTS

Whereas a lock-in keeps someone in a space or prevents an action until the desired operations have been done, a lockout prevents someone from entering a space that is dangerous, or prevents an event from occurring. A good example of a lockout is found in stairways of public buildings, at least in the United States (Figure 4.7). In cases of fire, people have a tendency to flee in panic, down the stairs, down, down, down, past the ground floor and into the basement, where they might be trapped. The solution (required by the fire laws) is not to allow simple passage from the ground floor to the basement.

Lockouts are usually used for safety reasons. Thus, small children are protected by baby locks on cabinet doors, covers for electric outlets, and specialized caps on containers for drugs and toxic substances. The pin that prevents a fire extinguisher from being activated until it is removed is a lockout forcing function to prevent accidental discharge.

Forcing functions can be a nuisance in normal usage. The result is that many people will deliberately disable the forcing function, thereby negating its safety feature. The clever designer has to minimize the nuisance value while retaining the safety feature of the forcing function that guards against the occasional tragedy. The gate in Figure 4.7 is a clever compromise: sufficient restraint to make people realize they are leaving the ground floor, but not enough of an impediment to normal behavior that people will prop open the gate.

Other useful devices make use of a forcing function. In some public restrooms, a pull-down shelf is placed inconveniently on the wall just behind the cubicle door, held in a vertical position by a spring. You lower the shelf to the horizontal position, and the weight of a package or handbag keeps it there. The shelf's position is a forcing function. When the shelf is lowered, it blocks the door fully. So to get out of the cubicle, you have to remove whatever is on the shelf and raise it out of the way. Clever design.

Conventions, Constraints, and Affordances

In Chapter 1 we learned of the distinctions between affordances, perceived affordances, and signifiers. Affordances refer to the potential actions that are possible, but these are easily discoverable only if they are perceivable: perceived affordances. It is the signifier component of the perceived affordance that allows people to determine the possible actions. But how does one go from the perception of an affordance to understanding the potential action? In many cases, through conventions.

A doorknob has the perceived affordance of graspability. But knowing that it is the doorknob that is used to open and close doors is learned: it is a cultural aspect of the design that knobs, handles, and bars, when placed on doors, are intended to enable the opening and shutting of those doors. The same devices on fixed walls would have a different interpretation: they might offer support, for example, but certainly not the possibility of opening the wall. The interpretation of a perceived affordance is a cultural convention.

CONVENTIONS ARE CULTURAL CONSTRAINTS

Conventions are a special kind of cultural constraint. For example, the means by which people eat is subject to strong cultural constraints and conventions. Different cultures use different eating utensils. Some eat primarily with the fingers and bread. Some use elaborate serving devices. The same is true of almost every aspect of behavior imaginable, from the clothes that are worn; to the way one addresses elders, equals, and inferiors; and even to the order in which people enter or exit a room. What is considered correct and proper in one culture may be considered impolite in another.

Although conventions provide valuable guidance for novel situations, their existence can make it difficult to enact change: consider the story of destination-control elevators.

WHEN CONVENTIONS CHANGE: THE CASE OF DESTINATION-CONTROL ELEVATORS

> *Operating the common elevator seems like a no-brainer. Press the button, get in the box, go up or down, get out. But we've been encountering and documenting an array of curious design variations on this simple interaction, raising the question: Why?* (From Portigal & Norvaisas, 2011.)

This quotation comes from two design professionals who were so offended by a change in the controls for an elevator system that they wrote an entire article of complaint.

What could possibly cause such an offense? Was it really bad design or, as the authors suggest, a completely unnecessary change to an otherwise satisfactory system? Here is what happened: the authors had encountered a new convention for elevators called "Elevator Destination Control." Many people (including me) consider it superior to the one we are all used to. Its major disadvantage is that it is different. It violates customary convention. Violations of convention can be very disturbing. Here is the history.

When "modern" elevators were first installed in buildings in the late 1800s, they always had a human operator who controlled the speed and direction of the elevator, stopped at the appropriate floors, and opened and shut the doors. People would enter the elevator, greet the operator, and state which floor they wished to travel to. When the elevators became automated, a similar convention was followed. People entered the elevator and told the elevator what floor they were traveling to by pushing the appropriately marked button inside the elevator.

This is a pretty inefficient way of doing things. Most of you have probably experienced a crowded elevator where every person seems to want to go to a different floor, which means a slow trip for the people going to the higher floors. A destination-control elevator system groups passengers, so that those going to the same floor are asked to use the same elevator and the passenger load is distributed to maximize efficiency. Although this kind of grouping is only sensible for buildings that have a large number of elevators, that would cover any large hotel, office, or apartment building.

In the traditional elevator, passengers stand in the elevator hallway and indicate whether they wish to travel up or down. When an elevator arrives going in the appropriate direction, they get in and use the keypad inside the elevator to indicate their destination floor. As a result, five people might get into the same elevator each wanting a different floor. With destination control, the destination keypads are located in the hallway outside the elevators and there are no keypads inside the elevators (Figure 4.8A and D). People are directed to whichever elevator will most efficiently reach their floor. Thus, if there were five people desiring elevators, they might be assigned to five different elevators. The result is faster trips for everyone, with a minimum of stops. Even if people are assigned to elevators that are not the next to arrive, they will get to their destinations faster than if they took earlier elevators.

Destination control was invented in 1985, but the first commercial installation didn't appear until 1990 (in Schindler elevators). Now, decades later, it is starting to appear more frequently as developers of tall buildings discover that destination control yields better service to passengers, or equal service with fewer elevators.

Horrors! As Figure 4.8D confirms, there are no controls inside the elevator to specify a floor. What if passengers change their minds

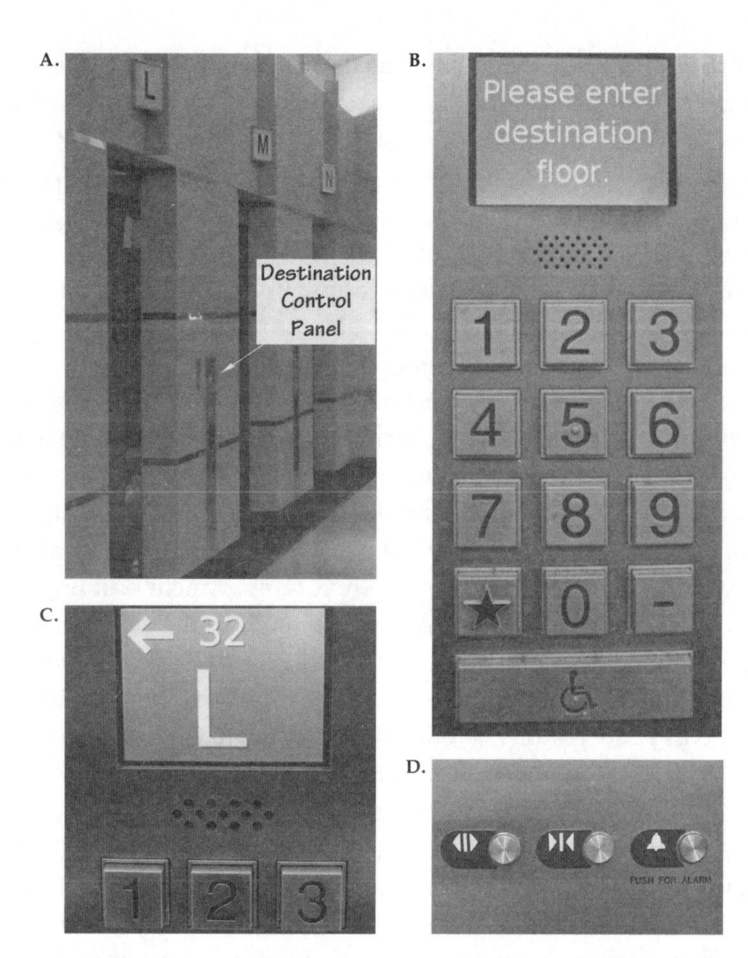

FIGURE 4.8. Destination-Control Elevators. In a destination-control system, the desired destination floor is entered into the control panel outside the elevators (A and B). After entering the destination floor into B, the display directs the traveler to the appropriate elevator, as shown in C, where "32" has been entered as the desired floor destination, and the person is directed to elevator "L" (the first elevator on the left, in A). There is no way to specify the floor from inside the elevator: Inside, the controls are only to open and shut the doors and an alarm (D). This is a much more efficient design, but confusing to people used to the more conventional system. (Photographs by the author.)

and wish to get off at a different floor? (Even my editor at Basic Books complained about this in a marginal note.) What then? What do you do in a regular elevator when you decide you really want to get off at the sixth floor just as the elevator passes the seventh floor? It's simple: just get off at the next stop and go to the destination control box in the elevator hall, and specify the intended floor.

PEOPLE'S RESPONSES TO CHANGES IN CONVENTIONS

People invariably object and complain whenever a new approach is introduced into an existing array of products and systems. Conventions are violated: new learning is required. The merits of the new system are irrelevant: it is the change that is upsetting. The destination control elevator is only one of many such examples. The metric system provides a powerful example of the difficulties in changing people's conventions.

The metric scale of measurement is superior to the English scale of units in almost every dimension: it is logical, easy to learn, and easy to use in computations. Today, over two centuries have passed since the metric system was developed by the French in the 1790s, yet three countries still resist its use: the United States, Liberia, and Myanmar. Even Great Britain has mostly switched, so the only major country left that uses the older English system of units is the United States. Why haven't we switched? The change is too upsetting for the people who have to learn the new system, and the initial cost of purchasing new tools and measuring devices seems excessive. The learning difficulties are nowhere as complex as purported, and the cost would be relatively small because the metric system is already in wide use, even in the United States.

Consistency in design is virtuous. It means that lessons learned with one system transfer readily to others. On the whole, consistency is to be followed. If a new way of doing things is only slightly better than the old, it is better to be consistent. But if there is to be a change, everybody has to change. Mixed systems are confusing to everyone. When a new way of doing things is vastly superior to another, then the merits of change outweigh the difficulty of change. Just because something is different does not mean it is bad. If we only kept to the old, we could never improve.

The Faucet: A Case History of Design

It may be hard to believe that an everyday water faucet could need an instruction manual. I saw one, this time at the meeting of the British Psychological Society in Sheffield, England. The participants were lodged in dormitories. Upon checking into Ranmoor House, each guest was given a pamphlet that provided useful information: where the churches were, the times of meals, the location of the post office, and how to work the taps (faucets). "The taps on the washhand basin are operated by pushing down gently."

When it was my turn to speak at the conference, I asked the audience about those taps. How many had trouble using them? Polite, restrained tittering from the audience. How many tried to turn the handle? A large show of hands. How many had to seek help? A few honest folks raised their hands. Afterward, one woman came up to me and said that she had given up and walked the halls until she found someone who could explain the taps to her. A simple sink, a simple-looking faucet. But it looks as if it should be turned, not pushed. If you want the faucet to be pushed, make it look as if it should be pushed. (This, of course, is similar to the problem I had emptying the water from the sink in my hotel, described in Chapter 1.)

Why is such a simple, standard item as a water faucet so difficult to get right? The person using a faucet cares about two things: water temperature and rate of flow. But water enters the faucet through two pipes, hot and cold. There is a conflict between the human need for temperature and flow and the physical structure of hot and cold.

There are several ways to deal with this:

- **Control both hot and cold water:** Two controls, one for hot water, the other cold.
- **Control only temperature:** One control, where rate of flow is fixed. Rotating the control from its fixed position turns on the water at

some predetermined rate of flow, with the temperature controlled by the knob position.

- **Control only amount:** One control, where temperature is fixed, with rate of flow controlled by the knob position.
- **On-off.** One control turns the water on and off. This is how gesture-controlled faucets work: moving the hand under or away from the spout turns the water on or off, at a fixed temperature and rate of flow.
- **Control temperature and rate of flow.** Use two separate controls, one for water temperature, the other for flow rate. (I have never encountered this solution.)
- **One control for temperature and rate:** Have one integrated control, where movement in one direction controls the temperature and movement in a different direction controls the amount.

Where there are two controls, one for hot water and one for cold, there are four mapping problems;

- Which knob controls the hot, which the cold?
- How do you change the temperature without affecting the rate of flow?
- How do you change the flow without affecting the temperature?
- Which direction increases water flow?

The mapping problems are solved through cultural conventions, or constraints. It is a worldwide convention that the left faucet should be hot; the right, cold. It is also a universal convention that screw threads are made to tighten with clockwise turning, loosen with counterclockwise. You turn off a faucet by tightening a screw thread (tightening a washer against its seat), thereby shutting off the flow of water. So clockwise turning shuts off the water, counterclockwise turns it on.

Unfortunately, the constraints do not always hold. Most of the English people I asked were not aware that left/hot, right/cold was a convention; it is violated too often to be considered a convention in England. But the convention isn't universal in the

United States, either. I once experienced shower controls that were placed vertically: Which one controlled the hot water, the top faucet or the bottom?

If the two faucet handles are round knobs, clockwise rotation of either should decrease volume. However, if each faucet has a single "blade" as its handle, then people don't think they are rotating the handles: they think that they are pushing or pulling. To maintain consistency, pulling either faucet should increase volume, even though this means rotating the left faucet counterclockwise and the right one clockwise. Although rotation direction is inconsistent, pulling and pushing is consistent, which is how people conceptualize their actions.

Alas, sometimes clever people are too clever for our good. Some well-meaning plumbing designers have decided that consistency should be ignored in favor of their own, private brand of psychology. The human body has mirror-image symmetry, say these pseudo-psychologists. So if the left hand moves clockwise, why, the right hand should move counterclockwise. Watch out, your plumber or architect may install a bathroom fixture whose clockwise rotation has a different result with the hot water than with the cold.

As you try to control the water temperature, soap running down over your eyes, groping to change the water control with one hand, soap or shampoo clutched in the other, you are guaranteed to get it wrong. If the water is too cold, the groping hand is just as likely to make the water colder as to make it scalding hot.

Whoever invented that mirror-image nonsense should be forced to take a shower. Yes, there is some logic to it. To be a bit fair to the inventor of the scheme, it works as long as you always use two hands to adjust both faucets simultaneously. It fails miserably, however, when one hand is used to alternate between the two controls. Then you cannot remember which direction does what. Once again, notice that this can be corrected without replacing the individual faucets: just replace the handles with blades. It is psychological perceptions that matter—the conceptual model—not physical consistency.

The operation of faucets needs to be standardized so that the psychological conceptual model of operation is the same for all types of faucets. With the traditional dual faucet controls for hot and cold water, the standards should state:

- When the handles are round, both should rotate in the same direction to change water volume.
- When the handles are single blades, both should be pulled to change water volume (which means rotating in opposite directions in the faucet itself).

Other configurations of handles are possible. Suppose the handles are mounted on a horizontal axis so that they rotate vertically. Then what? Would the answer differ for single blade handles and round ones? I leave this as an exercise for the reader.

What about the evaluation problem? Feedback in the use of most faucets is rapid and direct, so turning them the wrong way is easy to discover and correct. The evaluate-action cycle is easy to traverse. As a result, the discrepancy from normal rules is often not noticed—unless you are in the shower and the feedback occurs when you scald or freeze yourself. When the faucets are far removed from the spout, as is the case where the faucets are located in the center of the bathtub but the spouts high on an end wall, the delay between turning the faucets and the change in temperature can be quite long: I once timed a shower control to take 5 seconds. This makes setting the temperature rather difficult. Turn the faucet the wrong way and then dance around inside the shower while the water is scalding hot or freezing cold, madly turning the faucet in what you hope is the correct direction, hoping the temperature will stabilize quickly. Here the problem comes from the properties of fluid flow—it takes time for water to travel the 2 meters or so of pipe that might connect the faucets with the spout—so it is not easily remedied. But the problem is exacerbated by poor design of the controls.

Now let's turn to the modern single-spout, single-control faucet. Technology to the rescue. Move the control one way, it adjusts temperature. Move it another, it adjusts volume. Hurrah!

We control exactly the variables of interest, and the mixing spout solves the evaluation problem.

Yes, these new faucets are beautiful. Sleek, elegant, prize winning. Unusable. They solved one set of problems only to create yet another. The mapping problems now predominate. The difficulty lies in a lack of standardization of the dimensions of control, and then, which direction of movement means what? Sometimes there is a knob that can be pushed or pulled, rotated clockwise or counterclockwise. But does the push or pull control volume or temperature? Is a pull more volume or less, hotter temperature or cooler? Sometimes there is a lever that moves side to side or forward and backward. Once again, which movement is volume, which temperature? And even then, which way is more (or hotter), which is less (or cooler)? The perceptually simple one-control faucet still has four mapping problems:

- What dimension of control affects the temperature?
- Which direction along that dimension means hotter?
- What dimension of control affects the rate of flow?
- Which direction along that dimension means more?

In the name of elegance, the moving parts sometimes meld invisibly into the faucet structure, making it nearly impossible even to find the controls, let alone figure out which way they move or what they control. And then, different faucet designs use different solutions. One-control faucets ought to be superior because they control the psychological variables of interest. But because of the lack of standardization and awkward design (to call it "awkward" is being kind), they frustrate many people so much that they tend to be disliked more than they are admired.

Bath and kitchen faucet design ought to be simple, but can violate many design principles, including:

- Visible affordances and signifiers
- Discoverability
- Immediacy of feedback

Finally, many violate the principle of desperation:

- If all else fails, standardize.

Standardization is indeed the fundamental principle of desperation: when no other solution appears possible, simply design everything the same way, so people only have to learn once. If all makers of faucets could agree on a standard set of motions to control amount and temperature (how about up and down to control amount—up meaning increase—and left and right to control temperature, left meaning hot?), then we could all learn the standards once, and forever afterward use the knowledge for every new faucet we encountered.

If you can't put the knowledge on the device (that is, knowledge in the world), then develop a cultural constraint: standardize what has to be kept in the head. And remember the lesson from faucet rotation on page 153: The standards should reflect the psychological conceptual models, not the physical mechanics.

Standards simplify life for everyone. At the same time, they tend to hinder future development. And, as discussed in Chapter 6, there are often difficult political struggles in finding common agreement. Nonetheless, when all else fails, standards are the way to proceed.

Using Sound as Signifiers

Sometimes everything that is needed cannot be made visible. Enter sound: sound can provide information available in no other way. Sound can tell us that things are working properly or that they need maintenance or repair. It can even save us from accidents. Consider the information provided by:

- The click when the bolt on a door slides home
- The tinny sound when a door doesn't shut right
- The roaring sound when a car muffler gets a hole
- The rattle when things aren't secured
- The whistle of a teakettle when the water boils

- The click when the toast pops up
- The increase in pitch when a vacuum cleaner gets clogged
- The indescribable change in sound when a complex piece of machinery starts to have problems

Many devices simply beep and burp. These are not naturalistic sounds; they do not convey hidden information. When used properly, a beep can assure you that you've pressed a button, but the sound is as annoying as informative. Sounds should be generated so as to give knowledge about the source. They should convey something about the actions that are taking place, actions that matter to the user but that would otherwise not be visible. The buzzes, clicks, and hums that you hear while a telephone call is being completed are one good example: take out those noises and you are less certain that the connection is being made.

Real, natural sound is as essential as visual information because sound tells us about things we can't see, and it does so while our eyes are occupied elsewhere. Natural sounds reflect the complex interaction of natural objects: the way one part moves against another; the material of which the parts are made—hollow or solid, metal or wood, soft or hard, rough or smooth. Sounds are generated when materials interact, and the sound tells us whether they are hitting, sliding, breaking, tearing, crumbling, or bouncing. Experienced mechanics can diagnosis the condition of machinery just by listening. When sounds are generated artificially, if intelligently created using a rich auditory spectrum, with care to provide the subtle cues that are informative without being annoying, they can be as useful as sounds in the real world.

Sound is tricky. It can annoy and distract as easily as it can aid. Sounds that at one's first encounter are pleasant or cute easily become annoying rather than useful. One of the virtues of sounds is that they can be detected even when attention is applied elsewhere. But this virtue is also a deficit, for sounds are often intrusive. Sounds are difficult to keep private unless the intensity is low or earphones are used. This means both that neighbors may be annoyed and that others can monitor your activities. The use of sound to convey knowledge is a powerful and important idea, but still in its infancy.

Just as the presence of sound can serve a useful role in providing feedback about events, the absence of sound can lead to the same kinds of difficulties we have already encountered from a lack of feedback. The absence of sound can mean an absence of knowledge, and if feedback from an action is expected to come from sound, silence can lead to problems.

WHEN SILENCE KILLS

It was a pleasant June day in Munich, Germany. I was picked up at my hotel and driven to the country with farmland on either side of the narrow, two-lane road. Occasional walkers strode by, and every so often a bicyclist passed. We parked the car on the shoulder of the road and joined a group of people looking up and down the road. "Okay, get ready," I was told. "Close your eyes and listen." I did so and about a minute later I heard a high-pitched whine, accompanied by a low humming sound: an automobile was approaching. As it came closer, I could hear tire noise. After the car had passed, I was asked my judgment of the sound. We repeated the exercise numerous times, and each time the sound was different. What was going on? We were evaluating sound designs for BMW's new electric vehicles.

Electric cars are extremely quiet. The only sounds they make come from the tires, the air, and occasionally, from the high-pitched whine of the electronics. Car lovers really like the silence. Pedestrians have mixed feelings, but the blind are greatly concerned. After all, the blind cross streets in traffic by relying upon the sounds of vehicles. That's how they know when it is safe to cross. And what is true for the blind might also be true for anyone stepping onto the street while distracted. If the vehicles don't make any sounds, they can kill. The United States National Highway Traffic Safety Administration determined that pedestrians are considerably more likely to be hit by hybrid or electric vehicles than by those that have an internal combustion engine. The greatest danger is when the hybrid or electric vehicles are moving slowly, when they are almost completely silent. The sounds of an automobile are important signifiers of its presence.

Adding sound to a vehicle to warn pedestrians is not a new idea. For many years, commercial trucks and construction equipment have had to make beeping sounds when backing up. Horns are required by law, presumably so that drivers can use them to alert pedestrians and other drivers when the need arises, although they are often used as a way of venting anger and rage instead. But adding a continuous sound to a normal vehicle because it would otherwise be too quiet, is a challenge.

What sound would you want? One group of blind people suggested putting some rocks into the hubcaps. I thought this was brilliant. The rocks would provide a natural set of cues, rich in meaning yet easy to interpret. The car would be quiet until the wheels started to turn. Then, the rocks would make natural, continuous scraping sounds at low speeds, change to the pitter-patter of falling stones at higher speeds, the frequency of the drops increasing with the speed of the car until the car was moving fast enough that the rocks would be frozen against the circumference of the rim, silent. Which is fine: the sounds are not needed for fast-moving vehicles because then the tire noise is audible. The lack of sound when the vehicle was not moving would be a problem, however.

The marketing divisions of automobile manufacturers thought that the addition of artificial sounds would be a wonderful branding opportunity, so each car brand or model should have its own unique sound that captured just the car personality the brand wished to convey. Porsche added loudspeakers to its electric car prototype to give it the same "throaty growl" as its gasoline-powered cars. Nissan wondered whether a hybrid automobile should sound like tweeting birds. Some manufacturers thought all cars should sound the same, with standardized sounds and sound levels, making it easier for everyone to learn how to interpret them. Some blind people thought they should sound like cars—you know, gasoline engines, following the old tradition that new technologies must always copy the old.

Skeuomorphic is the technical term for incorporating old, familiar ideas into new technologies, even though they no longer play a functional role. Skeuomorphic designs are often comfortable for traditionalists, and indeed the history of technology shows that new technologies and materials often slavishly imitate the old for no apparent reason except that is what people know how to do. Early automobiles looked like horse-driven carriages without the horses (which is also why they were called horseless carriages); early plastics were designed to look like wood; folders in computer file systems often look the same as paper folders, complete with tabs. One way of overcoming the fear of the new is to make it look like the old. This practice is decried by design purists, but in fact, it has its benefits in easing the transition from the old to the new. It gives comfort and makes learning easier. Existing conceptual models need only be modified rather than replaced. Eventually, new forms emerge that have no relationship to the old, but the skeuomorphic designs probably helped the transition.

When it came to deciding what sounds the new silent automobiles should generate, those who wanted differentiation ruled the day, yet everyone also agreed that there had to be some standards. It should be possible to determine that the sound is coming from an automobile, to identify its location, direction, and speed. No sound would be necessary once the car was going fast enough, in part because tire noise would be sufficient. Some standardization would be required, although with a lot of leeway. International standards committees started their procedures. Various countries, unhappy with the normally glacial speed of standards agreements and under pressure from their communities, started drafting legislation. Companies scurried to develop appropriate sounds, hiring experts in psychoacoustics, psychologists, and Hollywood sound designers.

The United States National Highway Traffic Safety Administration issued a set of principles along with a detailed list of requirements, including sound levels, spectra, and other criteria. The full document is 248 pages. The document states:

> *This standard will ensure that blind, visually-impaired, and other pedestrians are able to detect and recognize nearby hybrid and electric vehicles by requiring that hybrid and electric vehicles emit sound that pedestrians will be able to hear in a range of ambient environments and contain acoustic signal content that pedestrians will recognize as being emitted from a vehicle. The proposed standard establishes minimum sound requirements for hybrid and electric vehicles when operating under 30 kilometers per hour (km/h) (18 mph), when the vehicle's starting system is activated but the vehicle is stationary, and when the vehicle is operating in reverse. The agency chose a crossover speed of 30 km/h because this was the speed at which the sound levels of the hybrid and electric vehicles measured by the agency approximated the sound levels produced by similar internal combustion engine vehicles.* (Department of Transportation, 2013.)

As I write this, sound designers are still experimenting. The automobile companies, lawmakers, and standards committees are still at work. Standards are not expected until 2014 or later, and then it will take considerable time to be deployed to the millions of vehicles across the world.

What principles should be used for the design sounds of electric vehicles (including hybrids)? The sounds have to meet several criteria:

- **Alerting.** The sound will indicate the presence of an electric vehicle.
- **Orientation.** The sound will make it possible to determine where the vehicle is located, a rough idea of its speed, and whether it is moving toward or away from the listener.
- **Lack of annoyance.** Because these sounds will be heard frequently even in light traffic and continually in heavy traffic, they must not be annoying. Note the contrast with sirens, horns, and backup signals, all of which are intended to be aggressive warnings. Such sounds are deliberately unpleasant, but because they are infrequent and for relatively short duration, they are acceptable. The challenge faced by electric vehicle sounds is to alert and orient, not annoy.

- **Standardization versus individualization.** Standardization is necessary to ensure that all electric vehicle sounds can readily be interpreted. If they vary too much, novel sounds might confuse the listener. Individualization has two functions: safety and marketing. From a safety point of view, if there were many vehicles present on the street, individualization would allow vehicles to be tracked. This is especially important at crowded intersections. From a marketing point of view, individualization can ensure that each brand of electric vehicle has its own unique characteristic, perhaps matching the quality of the sound to the brand image.

Stand still on a street corner and listen carefully to the vehicles around you. Listen to the silent bicycles and to the artificial sounds of electric cars. Do the cars meet the criteria? After years of trying to make cars run more quietly, who would have thought that one day we would spend years of effort and tens of millions of dollars to add sound?

HUMAN ERROR? NO, BAD DESIGN

Most industrial accidents are caused by human error: estimates range between 75 and 95 percent. How is it that so many people are so incompetent? Answer: They aren't. It's a design problem.

If the number of accidents blamed upon human error were 1 to 5 percent, I might believe that people were at fault. But when the percentage is so high, then clearly other factors must be involved. When something happens this frequently, there must be another underlying factor.

When a bridge collapses, we analyze the incident to find the causes of the collapse and reformulate the design rules to ensure that form of accident will never happen again. When we discover that electronic equipment is malfunctioning because it is responding to unavoidable electrical noise, we redesign the circuits to be more tolerant of the noise. But when an accident is thought to be caused by people, we blame them and then continue to do things just as we have always done.

Physical limitations are well understood by designers; mental limitations are greatly misunderstood. We should treat all failures in the same way: find the fundamental causes and redesign the system so that these can no longer lead to problems. We design equipment that requires people to be fully alert and attentive for hours, or to remember archaic, confusing procedures even if they are only used infrequently, sometimes only once in a lifetime. We put people in boring environments with nothing to do for hours on end, until suddenly they must respond quickly and accurately. Or we subject them to complex, high-workload environments, where they are continually interrupted while having to do multiple tasks simultaneously. Then we wonder why there is failure.

Even worse is that when I talk to the designers and administrators of these systems, they admit that they too have nodded off while supposedly working. Some even admit to falling asleep for an instant while driving. They admit to turning the wrong stove burners on or off in their homes, and to other small but significant errors. Yet when their workers do this, they blame them for "human error." And when employees or customers have similar issues, they are blamed for not following the directions properly, or for not being fully alert and attentive.

Understanding Why There Is Error

Error occurs for many reasons. The most common is in the nature of the tasks and procedures that require people to behave in unnatural ways—staying alert for hours at a time, providing precise, accurate control specifications, all the while multitasking, doing several things at once, and subjected to multiple interfering activities. Interruptions are a common reason for error, not helped by designs and procedures that assume full, dedicated attention yet that do not make it easy to resume operations after an interruption. And finally, perhaps the worst culprit of all, is the attitude of people toward errors.

When an error causes a financial loss or, worse, leads to an injury or death, a special committee is convened to investigate the cause and, almost without fail, guilty people are found. The next step is to blame and punish them with a monetary fine, or by firing or jailing them. Sometimes a lesser punishment is proclaimed: make the guilty parties go through more training. Blame and punish; blame and train. The investigations and resulting punishments feel good: "We caught the culprit." But it doesn't cure the problem: the same error will occur over and over again. Instead, when an error happens, we should determine why, then redesign the product or the procedures being followed so that it will never occur again or, if it does, so that it will have minimal impact.

ROOT CAUSE ANALYSIS

Root cause analysis is the name of the game: investigate the accident until the single, underlying cause is found. What this ought to mean is that when people have indeed made erroneous decisions or actions, we should determine what caused them to err. This is what root cause analysis ought to be about. Alas, all too often it stops once a person is found to have acted inappropriately.

Trying to find the cause of an accident sounds good but it is flawed for two reasons. First, most accidents do not have a single cause: there are usually multiple things that went wrong, multiple events that, had any one of them not occurred, would have prevented the accident. This is what James Reason, the noted British authority on human error, has called the "Swiss cheese model of accidents" (shown in Figure 5.3 of this chapter on page 208, and discussed in more detail there).

Second, why does the root cause analysis stop as soon as a human error is found? If a machine stops working, we don't stop the analysis when we discover a broken part. Instead, we ask: "Why did the part break? Was it an inferior part? Were the required specifications too low? Did something apply too high a load on the part?" We keep asking questions until we are satisfied that we understand the reasons for the failure: then we set out to remedy them. We should do the same thing when we find human error: We should discover what led to the error. When root cause analysis discovers a human error in the chain, its work has just begun: now we apply the analysis to understand why the error occurred, and what can be done to prevent it.

One of the most sophisticated airplanes in the world is the US Air Force's F-22. However, it has been involved in a number of accidents, and pilots have complained that they suffered oxygen deprivation (hypoxia). In 2010, a crash destroyed an F-22 and killed the pilot. The Air Force investigation board studied the incident and two years later, in 2012, released a report that blamed the accident on pilot error: "failure to recognize and initiate a timely dive recovery due to channelized attention, breakdown of visual scan and unrecognized spatial distortion."

In 2013, the Inspector General's office of the US Department of Defense reviewed the Air Force's findings, disagreeing with the assessment. In my opinion, this time a proper root cause analysis was done. The Inspector General asked "why sudden incapacitation or unconsciousness was not considered a contributory factor." The Air Force, to nobody's surprise, disagreed with the criticism. They argued that they had done a thorough review and that their conclusion "was supported by clear and convincing evidence." Their only fault was that the report "could have been more clearly written."

It is only slightly unfair to parody the two reports this way:

> ***Air Force:*** It was pilot error—the pilot failed to take corrective action.
>
> ***Inspector General:*** That's because the pilot was probably unconscious.
>
> ***Air Force:*** So you agree, the pilot failed to correct the problem.

THE FIVE WHYS

Root cause analysis is intended to determine the underlying cause of an incident, not the proximate cause. The Japanese have long followed a procedure for getting at root causes that they call the "Five Whys," originally developed by Sakichi Toyoda and used by the Toyota Motor Company as part of the Toyota Production System for improving quality. Today it is widely deployed. Basically, it means that when searching for the reason, even after you have found one, do not stop: ask why that was the case. And then ask why again. Keep asking until you have uncovered the true underlying causes. Does it take exactly five? No, but calling the procedure "Five Whys" emphasizes the need to keep going even after a reason has been found. Consider how this might be applied to the analysis of the F-22 crash:

Five Whys

Question	Answer
Q1: Why did the plane crash?	Because it was in an uncontrolled dive.
Q2: Why didn't the pilot recover from the dive?	Because the pilot failed to initiate a timely recovery.
Q3: Why was that?	Because he might have been unconscious (or oxygen deprived).
Q4: Why was that?	We don't know. We need to find out.
Etc.	

The Five Whys of this example are only a partial analysis. For example, we need to know why the plane was in a dive (the report explains this, but it is too technical to go into here; suffice it to say that it, too, suggests that the dive was related to a possible oxygen deprivation).

The Five Whys do not guarantee success. The question *why* is ambiguous and can lead to different answers by different investigators. There is still a tendency to stop too soon, perhaps when the limit of the investigator's understanding has been reached. It also tends to emphasize the need to find a single cause for an incident, whereas most complex events have multiple, complex causal factors. Nonetheless, it is a powerful technique.

The tendency to stop seeking reasons as soon as a human error has been found is widespread. I once reviewed a number of accidents in which highly trained workers at an electric utility company had been electrocuted when they contacted or came too close to the high-voltage lines they were servicing. All the investigating committees found the workers to be at fault, something even the workers (those who had survived) did not dispute. But when the committees were investigating the complex causes of the incidents, why did they stop once they found a human error? Why didn't they keep going to find out why the error had occurred, what circumstances had led to it, and then, why those circumstances had happened? The committees never went far enough to find the deeper, root causes of the accidents. Nor did they consider redesigning the systems and procedures to make the incidents either impossible or far less likely. When people err, change the system so that type of error will be reduced or eliminated. When complete elimination is not possible, redesign to reduce the impact.

It wasn't difficult for me to suggest simple changes to procedures that would have prevented most of the incidents at the utility company. It had never occurred to the committee to think of this. The problem is that to have followed my recommendations would have meant changing the culture from an attitude among the field workers that "We are supermen: we can solve any problem, repair the most complex outage. We do not make errors." It is not possible to eliminate human error if it is thought of as a personal failure rather than as a sign of poor design of procedures or equipment. My report to the company executives was received politely. I was even thanked. Several years later I contacted a friend at the company and asked what changes they had made. "No changes," he said. "And we are still injuring people."

One big problem is that the natural tendency to blame someone for an error is even shared by those who made the error, who often agree that it was their fault. People do tend to blame themselves when they do something that, after the fact, seems inexcusable. "I knew better," is a common comment by those who have erred. But when someone says, "It was my fault, I knew better," this is not a valid analysis of the problem. That doesn't help prevent its recurrence. When many people all have the same problem, shouldn't another cause be found? If the system lets you make the error, it is badly designed. And if the system induces you to make the error, then it is really badly designed. When I turn on the wrong stove burner, it is not due to my lack of knowledge: it is due to poor mapping between controls and burners. Teaching me the relationship will not stop the error from recurring: redesigning the stove will.

We can't fix problems unless people admit they exist. When we blame people, it is then difficult to convince organizations to restructure the design to eliminate these problems. After all, if a person is at fault, replace the person. But seldom is this the case: usually the system, the procedures, and social pressures have led to the problems, and the problems won't be fixed without addressing all of these factors.

Why do people err? Because the designs focus upon the requirements of the system and the machines, and not upon the requirements of people. Most machines require precise commands and guidance, forcing people to enter numerical information perfectly. But people aren't very good at great precision. We frequently make errors when asked to type or write sequences of numbers or letters. This is well known: so why are machines still being designed that require such great precision, where pressing the wrong key can lead to horrendous results?

People are creative, constructive, exploratory beings. We are particularly good at novelty, at creating new ways of doing things, and at seeing new opportunities. Dull, repetitive, precise requirements fight against these traits. We are alert to changes in the environment, noticing new things, and then thinking about them and their implications. These are virtues, but they get turned into negative features when we are forced to serve machines. Then we are punished for lapses in attention, for deviating from the tightly prescribed routines.

A major cause of error is time stress. Time is often critical, especially in such places as manufacturing or chemical processing plants and hospitals. But even everyday tasks can have time pressures. Add environmental factors, such as poor weather or heavy traffic, and the time stresses increase. In commercial establishments, there is strong pressure not to slow the processes, because doing so would inconvenience many, lead to significant loss of money, and, in a hospital, possibly decrease the quality of patient care. There is a lot of pressure to push ahead with the work even when an outside observer would say it was dangerous to do so. In many industries, if the operators actually obeyed all the procedures, the work would never get done. So we push the boundaries: we stay up far longer than is natural. We try to do too many tasks at the same time. We drive faster than is safe. Most of the time we manage okay. We might even be rewarded and praised for our heroic efforts. But when things go wrong and we fail, then this same behavior is blamed and punished.

Deliberate Violations

Errors are not the only type of human failures. Sometimes people knowingly take risks. When the outcome is positive, they are often rewarded. When the result is negative, they might be punished. But how do we classify these deliberate violations of known, proper behavior? In the error literature, they tend to be ignored. In the accident literature, they are an important component.

Deliberate deviations play an important role in many accidents. They are defined as cases where people intentionally violate procedures and regulations. Why do they happen? Well, almost every one of us has probably deliberately violated laws, rules, or even our own best judgment at times. Ever go faster than the speed limit? Drive too fast in the snow or rain? Agree to do some hazardous act, even while privately thinking it foolhardy to do so?

In many industries, the rules are written more with a goal toward legal compliance than with an understanding of the work requirements. As a result, if workers followed the rules, they couldn't get their jobs done. Do you sometimes prop open locked doors? Drive with too little sleep? Work with co-workers even though you are ill (and might therefore be infectious)?

Routine violations occur when noncompliance is so frequent that it is ignored. Situational violations occur when there are special circumstances (example: going through a red light "because no other cars were visible and I was late"). In some cases, the only way to complete a job might be to violate a rule or procedure.

A major cause of violations is inappropriate rules or procedures that not only invite violation but encourage it. Without the violations, the work could not be done. Worse, when employees feel it necessary to violate the rules in order to get the job done and, as a result, succeed, they will probably be congratulated and rewarded. This, of course, unwittingly rewards noncompliance. Cultures that encourage and commend violations set poor role models.

Although violations are a form of error, these are organizational and societal errors, important but outside the scope of the design of everyday things. The human error examined here is unintentional: deliberate violations, by definition, are intentional deviations that are known to be risky, with the potential of doing harm.

Two Types of Errors: Slips and Mistakes

Many years ago, the British psychologist James Reason and I developed a general classification of human error. We divided human error into two major categories: slips and mistakes (Figure 5.1). This classification has proved to be of value for both theory and practice. It is widely used in the study of error in such diverse areas as industrial and aviation accidents, and medical errors. The discussion gets a little technical, so I have kept technicalities to a minimum. This topic is of extreme importance to design, so stick with it.

DEFINITIONS: ERRORS, SLIPS, AND MISTAKES

Human error is defined as any deviance from "appropriate" behavior. The word *appropriate* is in quotes because in many circumstances, the appropriate behavior is not known or is only deter-

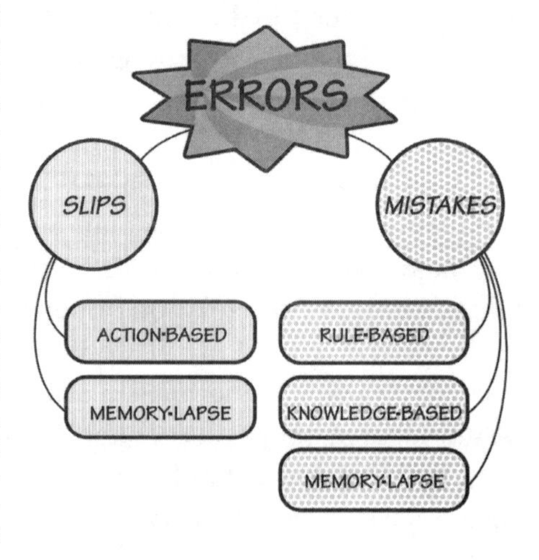

FIGURE 5.1. Classification of Errors. Errors have two major forms. Slips occur when the goal is correct, but the required actions are not done properly: the execution is flawed. Mistakes occur when the goal or plan is wrong. Slips and mistakes can be further divided based upon their underlying causes. Memory lapses can lead to either slips or mistakes, depending upon whether the memory failure was at the highest level of cognition (mistakes) or at lower (subconscious) levels (slips). Although deliberate violations of procedures are clearly inappropriate behaviors that often lead to accidents, these are not considered as errors (see discussion in text).

mined after the fact. But still, error is defined as deviance from the generally accepted correct or appropriate behavior.

Error is the general term for all wrong actions. There are two major classes of error: *slips* and *mistakes*, as shown in Figure 5.1; slips are further divided into two major classes and mistakes into three. These categories of errors all have different implications for design. I now turn to a more detailed look at these classes of errors and their design implications.

SLIPS

A slip occurs when a person intends to do one action and ends up doing something else. With a slip, the action performed is not the same as the action that was intended.

There are two major classes of slips: *action-based* and *memory-lapse.* In action-based slips, the wrong action is performed. In lapses, memory fails, so the intended action is not done or its results not evaluated. Action-based slips and memory lapses can be further classified according to their causes.

Example of an action-based slip. I poured some milk into my coffee and then put the coffee cup into the refrigerator. This is the correct action applied to the wrong object.

Example of a memory-lapse slip. I forget to turn off the gas burner on my stove after cooking dinner.

MISTAKES

A mistake occurs when the wrong goal is established or the wrong plan is formed. From that point on, even if the actions are executed properly they are part of the error, because the actions themselves are inappropriate—they are part of the wrong plan. With a mistake, the action that is performed matches the plan: it is the plan that is wrong.

Mistakes have three major classes: *rule-based, knowledge-based,* and *memory-lapse.* In a rule-based mistake, the person has appropriately diagnosed the situation, but then decided upon an erroneous course of action: the wrong rule is being followed. In a knowledge-based mistake, the problem is misdiagnosed because of erroneous or incomplete knowledge. Memory-lapse mistakes take place when there is forgetting at the stages of goals, plans, or evaluation. Two of the mistakes leading to the "Gimli Glider" Boeing 767 emergency landing were:

Example of knowledge-based mistake. Weight of fuel was computed in pounds instead of kilograms.

Example of memory-lapse mistake. A mechanic failed to complete troubleshooting because of distraction.

ERROR AND THE SEVEN STAGES OF ACTION

Errors can be understood through reference to the seven stages of the action cycle of Chapter 2 (Figure 5.2). Mistakes are errors in setting the goal or plan, and in comparing results with expectations—the higher levels of cognition. Slips happen in the execution of a plan, or in the perception or interpretation of the outcome—the lower stages. Memory lapses can happen at any of the eight transitions between stages, shown by the X's in Figure 5.2B. A memory lapse at one of these transitions stops the action cycle from proceeding, and so the desired action is not completed.

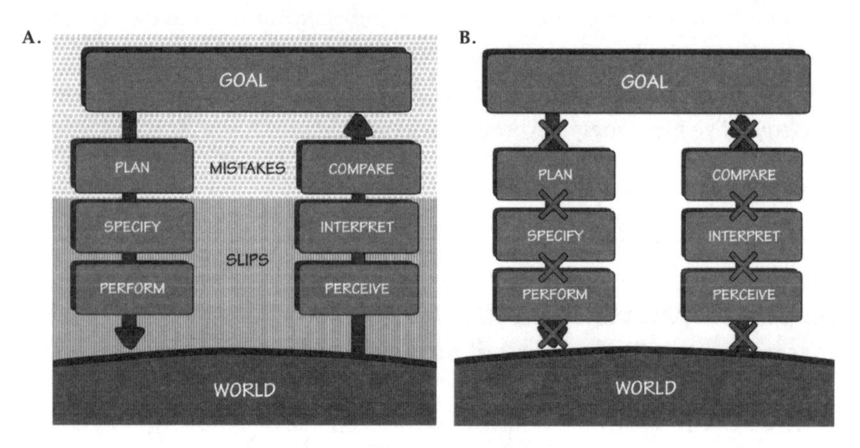

FIGURE 5.2. **Where Slips and Mistakes Originate in the Action Cycle.** Figure A shows that action slips come from the bottom four stages of the action cycle and mistakes from the top three stages. Memory lapses impact the transitions between stages (shown by the X's in Figure B). Memory lapses at the higher levels lead to mistakes, and lapses at the lower levels lead to slips.

Slips are the result of subconscious actions getting waylaid en route. Mistakes result from conscious deliberations. The same processes that make us creative and insightful by allowing us to see relationships between apparently unrelated things, that let us leap to correct conclusions on the basis of partial or even faulty evidence, also lead to mistakes. Our ability to generalize from small amounts of information helps tremendously in new situations; but sometimes we generalize too rapidly, classifying a new situation as similar to an old one when, in fact, there are significant discrepancies. This leads to mistakes that can be difficult to discover, let alone eliminate.

The Classification of Slips

A colleague reported that he went to his car to drive to work. As he drove away, he realized that he had forgotten his briefcase, so he turned around and went back. He stopped the car, turned off the engine, and unbuckled his wristwatch. Yes, his wristwatch, instead of his seatbelt.

The story illustrates both a memory-lapse slip and an action slip. The forgetting of the briefcase is a memory-lapse slip. The unbuckling of the wristwatch is an action slip, in this case a combination of description-similarity and capture error (described later in this chapter).

Most everyday errors are slips. Intending to do one action, you find yourself doing another. When a person says something clearly and distinctly to you, you "hear" something quite different. The study of slips is the study of the psychology of everyday errors—what Freud called "the psychopathology of everyday life." Freud believed that slips have hidden, dark meanings, but most are accounted for by rather simple mental mechanisms.

An interesting property of slips is that, paradoxically, they tend to occur more frequently to skilled people than to novices. Why? Because slips often result from a lack of attention to the task. Skilled people—experts—tend to perform tasks automatically, under subconscious control. Novices have to pay considerable conscious attention, resulting in a relatively low occurrence of slips.

Some slips result from the similarities of actions. Or an event in the world may automatically trigger an action. Sometimes our thoughts and actions may remind us of unintended actions, which we then perform. There are numerous different kinds of action slips, categorized by the underlying mechanisms that give rise to them. The three most relevant to design are:

- capture slips
- description-similarity slips
- mode errors

CAPTURE SLIPS

> *I was using a copying machine, and I was counting the pages. I found myself counting, "1, 2, 3, 4, 5, 6, 7, 8, 9, 10, Jack, Queen, King." I had been playing cards recently.*

The capture slip is defined as the situation where, instead of the desired activity, a more frequently or recently performed one gets done instead: it captures the activity. Capture errors require that part of the action sequences involved in the two activities be identical, with one sequence being far more familiar than the other. After doing the identical part, the more frequent or more recent activity continues, and the intended one does not get done. Seldom, if ever, does the unfamiliar sequence capture the familiar one. All that is needed is a lapse of attention to the desired action at the critical junction when the identical portions of the sequences diverge into the two different activities. Capture errors are, therefore, partial memory-lapse errors. Interestingly, capture errors are more prevalent in experienced skilled people than in beginners, in part because the experienced person has automated the required actions and may not be paying conscious attention when the intended action deviates from the more frequent one.

Designers need to avoid procedures that have identical opening steps but then diverge. The more experienced the workers, the more likely they are to fall prey to capture. Whenever possible, sequences should be designed to differ from the very start.

DESCRIPTION-SIMILARITY SLIPS

A former student reported that one day he came home from jogging, took off his sweaty shirt, and rolled it up in a ball, intending to throw it in the laundry basket. Instead he threw it in the toilet. (It wasn't poor aim: the laundry basket and toilet were in different rooms.)

In the slip known as a description-similarity slip, the error is to act upon an item similar to the target. This happens when the description of the target is sufficiently vague. Much as we saw in Chapter 3, Figure 3.1, where people had difficulty distinguishing among different images of money because their internal descriptions did not have sufficient discriminating information, the same thing can happen to us, especially when we are tired, stressed, or overloaded. In the example that opened this section, both the laundry basket and the toilet bowl are containers, and if the description of the target was sufficiently ambiguous, such as "a large enough container," the slip could be triggered.

Remember the discussion in Chapter 3 that most objects don't need precise descriptions, simply enough precision to distinguish the desired target from alternatives. This means that a description that usually suffices may fail when the situation changes so that multiple similar items now match the description. Description-similarity errors result in performing the correct action on the wrong object. Obviously, the more the wrong and right objects have in common, the more likely the errors are to occur. Similarly, the more objects present at the same time, the more likely the error.

Designers need to ensure that controls and displays for different purposes are significantly different from one another. A lineup of identical-looking switches or displays is very apt to lead to description-similarity error. In the design of airplane cockpits, many controls are shape coded so that they both look and feel different from one another: the throttle levers are different from the flap levers (which might look and feel like a wing flap), which are different from the landing gear control (which might look and feel like a wheel).

MEMORY-LAPSE SLIPS

Errors caused by memory failures are common. Consider these examples:

- Making copies of a document, walking off with the copy, but leaving the original inside the machine.
- Forgetting a child. This error has numerous examples, such as leaving a child behind at a rest stop during a car trip, or in the dressing room of a department store, or a new mother forgetting her one-month-old and having to go to the police for help in finding the baby.
- Losing a pen because it was taken out to write something, then put down while doing some other task. The pen is forgotten in the activities of putting away a checkbook, picking up goods, talking to a salesperson or friends, and so on. Or the reverse: borrowing a pen, using it, and then putting it away in your pocket or purse, even though it is someone else's (this is also a capture error).
- Using a bank or credit card to withdraw money from an automatic teller machine, then walking off without the card, is such a frequent error that many machines now have a forcing function: the card must be removed before the money will be delivered. Of course, it is then possible to walk off without the money, but this is less likely than forgetting the card because money is the goal of using the machine.

Memory lapses are common causes of error. They can lead to several kinds of errors: failing to do all of the steps of a procedure; repeating steps; forgetting the outcome of an action; or forgetting the goal or plan, thereby causing the action to be stopped.

The immediate cause of most memory-lapse failures is interruptions, events that intervene between the time an action is decided upon and the time it is completed. Quite often the interference comes from the machines we are using: the many steps required between the start and finish of the operations can overload the capacity of short-term or working memory.

There are several ways to combat memory-lapse errors. One is to minimize the number of steps; another, to provide vivid reminders of steps that need to be completed. A superior method is to use the forcing function of Chapter 4. For example, automated teller machines often require removal of the bank card before delivering the requested money: this prevents forgetting the bank card, capitalizing on the fact that people seldom forget the goal of the activity, in this case the money. With pens, the solution is simply to prevent their removal, perhaps by chaining public pens to the counter. Not all memory-lapse errors lend themselves to simple solutions. In many cases the interruptions come from outside the system, where the designer has no control.

MODE-ERROR SLIPS

A mode error occurs when a device has different states in which the same controls have different meanings: we call these states *modes.* Mode errors are inevitable in anything that has more possible actions than it has controls or displays; that is, the controls mean different things in the different modes. This is unavoidable as we add more and more functions to our devices.

Ever turn off the wrong device in your home entertainment system? This happens when one control is used for multiple purposes. In the home, this is simply frustrating. In industry, the confusion that results when operators believe the system to be in one mode, when in reality it is in another, has resulted in serious accidents and loss of life.

It is tempting to save money and space by having a single control serve multiple purposes. Suppose there are ten different functions on a device. Instead of using ten separate knobs or switches—which would take considerable space, add extra cost, and appear intimidatingly complex, why not use just two controls, one to select the function, the other to set the function to the desired condition? Although the resulting design appears quite simple and easy to use, this apparent simplicity masks the underlying complexity of use. The operator must always be completely aware of the mode, of what function is active. Alas, the prevalence of mode errors shows this assumption to be false. Yes, if I select a mode and then immediately adjust the parameters, I am not apt to be confused about the state. But what if I select the mode and then get interrupted by other events? Or if the mode is maintained for considerable periods? Or, as in the case of the Airbus accident discussed below, the two modes being selected are very similar in control and function, but have different operating characteristics, which means that the resulting mode error is difficult to discover? Sometimes the use of modes is justifiable, such as the need to put many controls and displays in a small, restricted space, but whatever the reason, modes are a common cause of confusion and error.

Alarm clocks often use the same controls and display for setting the time of day and the time the alarm should go off, and many of us have thereby set one when we meant the other. Similarly, when time is displayed on a twelve-hour scale, it is easy to set the alarm to go off at seven A.M. only later to discover that the alarm had been set for seven P.M. The use of "A.M." and "P.M." to distinguish times before and after noon is a common source of confusion and error, hence the common use of 24-hour time specification throughout most of the world (the major exceptions being North America, Australia, India, and the Philippines). Watches with multiple functions have similar problems, in this case required because of the small amount of space available for controls and displays. Modes exist in most computer programs, in our cell phones, and in the automatic controls of commercial aircraft. A number of serious accidents in commercial aviation can be attributed to mode errors, especially in aircraft that use automatic systems (which have a large number of complex modes). As automobiles become more complex, with the dashboard controls for driving, heating and air-conditioning, entertainment, and navigation, modes are increasingly common.

An accident with an Airbus airplane illustrates the problem. The flight control equipment (often referred to as the automatic pilot) had two modes, one for controlling vertical speed, the other for controlling the flight path's angle of descent. In one case, when the pilots were attempting to land, the pilots thought that they were controlling the angle of descent, whereas they had accidentally selected the mode that controlled speed of descent. The number (–3.3) that was entered into the system to represent an appropriate angle (–3.3°) was too steep a rate of descent when interpreted as vertical speed (–3,300 feet/minute: –3.3° would only be –800 feet/minute). This mode confusion contributed to the resulting fatal accident. After a detailed study of the accident, Airbus changed the display on the instrument so that vertical speed would always be displayed with a four-digit number and angle with two digits, thus reducing the chance of confusion.

Mode error is really design error. Mode errors are especially likely where the equipment does not make the mode visible, so the user is expected to remember what mode has been established, sometimes hours earlier, during which time many intervening events might have occurred. Designers must try to avoid modes, but if they are necessary, the equipment must make it obvious which mode is invoked. Once again, designers must always compensate for interfering activities.

The Classification of Mistakes

Mistakes result from the choice of inappropriate goals and plans or from faulty comparison of the outcome with the goals during evaluation. In mistakes, a person makes a poor decision, misclassifies a situation, or fails to take all the relevant factors into account. Many mistakes arise from the vagaries of human thought, often because people tend to rely upon remembered experiences rather than on more systematic analysis. We make decisions based upon what is in our memory. But as discussed in Chapter 3, retrieval from long-term memory is actually a reconstruction rather than an accurate record. As a result, it is subject to numerous biases. Among other things, our memories tend to be biased toward overgeneralization of the commonplace and overemphasis of the discrepant.

The Danish engineer Jens Rasmussen distinguished among three modes of behavior: skill-based, rule-based, and knowledge-based. This three-level classification scheme provides a practical tool that has found wide acceptance in applied areas, such as the design of many industrial systems. Skill-based behavior occurs when workers are extremely expert at their jobs, so they can do the everyday, routine tasks with little or no thought or conscious attention. The most common form of errors in skill-based behavior is slips.

Rule-based behavior occurs when the normal routine is no longer applicable but the new situation is one that is known, so there is already a well-prescribed course of action: a rule. Rules simply might be learned behaviors from previous experiences, but includes formal procedures prescribed in courses and manuals, usually in the form of "if-then" statements, such as, "*If* the engine will not start, *then* do [the appropriate action]." Errors with rule-based behavior can be either a mistake or a slip. If the wrong rule is selected, this would be a mistake. If the error occurs during the execution of the rule, it is most likely a slip.

Knowledge-based procedures occur when unfamiliar events occur, where neither existing skills nor rules apply. In this case, there must be considerable reasoning and problem-solving. Plans might be developed, tested, and then used or modified. Here, conceptual models are essential in guiding development of the plan and interpretation of the situation.

In both rule-based and knowledge-based situations, the most serious mistakes occur when the situation is misdiagnosed. As a result, an inappropriate rule is executed, or in the case of knowledge-based problems, the effort is addressed to solving the wrong problem. In addition, with misdiagnosis of the problem comes misinterpretation of the environment, as well as faulty comparisons of the current state with expectations. These kinds of mistakes can be very difficult to detect and correct.

RULE-BASED MISTAKES

When new procedures have to be invoked or when simple problems arise, we can characterize the actions of skilled people as rule-based. Some rules come from experience; others are formal procedures in manuals or rulebooks, or even less formal guides, such as cookbooks for food preparation. In either case, all we must do is identify the situation, select the proper rule, and then follow it.

When driving, behavior follows well-learned rules. Is the light red? If so, stop the car. Wish to turn left? Signal the intention to turn and move as far left as legally permitted: slow the vehicle and wait for a safe break in traffic, all the while following the traffic rules and relevant signs and lights.

Rule-based mistakes occur in multiple ways:

- The situation is mistakenly interpreted, thereby invoking the wrong goal or plan, leading to following an inappropriate rule.
- The correct rule is invoked, but the rule itself is faulty, either because it was formulated improperly or because conditions are different than assumed by the rule or through incomplete knowledge used to determine the rule. All of these lead to knowledge-based mistakes.
- The correct rule is invoked, but the outcome is incorrectly evaluated. This error in evaluation, usually rule- or knowledge-based itself, can lead to further problems as the action cycle continues.

Example 1: In 2013, at the Kiss nightclub in Santa Maria, Brazil, pyrotechnics used by the band ignited a fire that killed over 230 people. The tragedy illustrates several mistakes. The band made a knowledge-based mistake when they used outdoor flares, which ignited the ceiling's acoustic tiles. The band thought the flares were safe. Many people rushed into the rest rooms, mistakenly thinking they were exits: they died. Early reports suggested that the guards, unaware of the fire, at first mistakenly blocked people from leaving the building. Why? Because nightclub attendees would sometimes leave without paying for their drinks.

The mistake was in devising a rule that did not take account of emergencies. A root cause analysis would reveal that the goal was to prevent inappropriate exit but still allow the doors to be used in an emergency. One solution is doors that trigger alarms when used, deterring people trying to sneak out, but allowing exit when needed.

Example 2: Turning the thermostat of an oven to its maximum temperature to get it to the proper cooking temperature faster is a mistake based upon a false conceptual model of the way the oven works. If the person wanders off and forgets to come back and check the oven temperature after a reasonable period (a memory-lapse slip), the improper high setting of the oven temperature can lead to an accident, possibly a fire.

Example 3: A driver, unaccustomed to anti-lock brakes, encounters an unexpected object in the road on a wet, rainy day. The driver applies full force to the brakes but the car skids, triggering the anti-lock brakes to rapidly turn the brakes on and off, as they are designed to do. The driver, feeling the vibrations, believes that it indicates malfunction and therefore lifts his foot off the brake pedal. In fact, the vibration is a signal that anti-lock brakes are working properly. The driver's misevaluation leads to the wrong behavior.

Rule-based mistakes are difficult to avoid and then difficult to detect. Once the situation has been classified, the selection of the appropriate rule is often straightforward. But what if the classification of the situation is wrong? This is difficult to discover because there is usually considerable evidence to support the erroneous classification of the situation and the choice of rule. In complex situations, the problem is too much information: information that both supports the decision and also contradicts it. In the face of time pressures to make a decision, it is difficult to know which evidence to consider, which to reject. People usually decide by taking the current situation and matching it with something that happened earlier. Although human memory is quite good at matching examples from the past with the present situation, this doesn't mean that the matching is accurate or appropriate. The matching is biased by recency, regularity, and uniqueness. Recent events are remembered far better than less recent ones. Frequent events are remembered through their regularities, and unique events are remembered because of their uniqueness. But suppose the current event is different from all that has been experienced before: people are still apt to find some match in memory to use as a guide. The same powers that make us so good at dealing with the common and the unique lead to severe error with novel events.

What is a designer to do? Provide as much guidance as possible to ensure that the current state of things is displayed in a coherent and easily interpreted format—ideally graphical. This is a difficult problem. All major decision makers worry about the complexity of real-world events, where the problem is often too much information, much of it contradictory. Often, decisions must be made quickly. Sometimes it isn't even clear that there is an incident or that a decision is actually being made.

Think of it like this. In your home, there are probably a number of broken or misbehaving items. There might be some burnt-out lights, or (in my home) a reading light that works fine for a little while, then goes out: we have to walk over and wiggle the fluorescent bulb. There might be a leaky faucet or other minor faults that you know about but are postponing action to remedy. Now consider a major process-control manufacturing plant (an oil refinery, a chemical plant, or a nuclear power plant). These have thousands, perhaps tens of thousands, of valves and gauges, displays and controls, and so on. Even the best of plants always has some faulty parts. The maintenance crews always have a list of items to take care of. With all the alarms that trigger when a problem arises, even though it might be minor, and all the everyday failures, how does one know which might be a significant indicator of a major problem? Every single one usually has a simple, rational explanation, so not making it an urgent item is a sensible decision. In fact, the maintenance crew simply adds it to a list. Most of the time, this is the correct decision. The one time in a thousand (or even, one time in a million) that the decision is wrong makes it the one they will be blamed for: how could they have missed such obvious signals?

Hindsight is always superior to foresight. When the accident investigation committee reviews the event that contributed to the problem, they know what actually happened, so it is easy for them to pick out which information was relevant, which was not. This is retrospective decision making. But when the incident was taking place, the people were probably overwhelmed with far too much irrelevant information and probably not a lot of relevant information. How were they to know which to attend to and which to ignore? Most of the time, experienced operators get things right. The one time they fail, the retrospective analysis is apt to condemn them for missing the obvious. Well, during the event, nothing may be obvious. I return to this topic later in the chapter.

You will face this while driving, while handling your finances, and while just going through your daily life. Most of the unusual incidents you read about are not relevant to you, so you can safely ignore them. Which things should be paid attention to, which should be ignored? Industry faces this problem all the time, as do governments. The intelligence communities are swamped with data. How do they decide which cases are serious? The public hears about their mistakes, but not about the far more frequent cases that they got right or about the times they ignored data as not being meaningful—and were correct to do so.

If every decision had to be questioned, nothing would ever get done. But if decisions are not questioned, there will be major mistakes—rarely, but often of substantial penalty.

The design challenge is to present the information about the state of the system (a device, vehicle, plant, or activities being monitored) in a way that is easy to assimilate and interpret, as well as to provide alternative explanations and interpretations. It is useful to question decisions, but impossible to do so if every action—or failure to act—requires close attention.

This is a difficult problem with no obvious solution.

KNOWLEDGE-BASED MISTAKES

Knowledge-based behavior takes place when the situation is novel enough that there are no skills or rules to cover it. In this case, a new procedure must be devised. Whereas skills and rules are controlled at the behavioral level of human processing and are therefore subconscious and automatic, knowledge-based behavior is controlled at the reflective level and is slow and conscious.

With knowledge-based behavior, people are consciously problem solving. They are in an unknown situation and do not have any available skills or rules that apply directly. Knowledge-based behavior is required either when a person encounters an unknown situation, perhaps being asked to use some novel equipment, or even when doing a familiar task and things go wrong, leading to a novel, uninterpretable state.

The best solution to knowledge-based situations is to be found in a good understanding of the situation, which in most cases also translates into an appropriate conceptual model. In complex cases, help is needed, and here is where good cooperative problem-solving skills and tools are required. Sometimes, good procedural manuals (paper or electronic) will do the job, especially if critical observations can be used to arrive at the relevant procedures to follow. A more powerful approach is to develop intelligent computer systems, using good search and appropriate reasoning techniques (artificial-intelligence decision-making and problem-solving). The difficulties here are in establishing the interaction of the people with the automation: human teams and automated systems have to be thought of as collaborative, cooperative systems. Instead, they are often built by assigning the tasks that machines can do to the machines and leaving the humans to do the rest. This usually means that machines do the parts that are easy for people, but when the problems become complex, which is precisely when people could use assistance, that is when the machines usually fail. (I discuss this problem extensively in *The Design of Future Things*.)

MEMORY-LAPSE MISTAKES

Memory lapses can lead to mistakes if the memory failure leads to forgetting the goal or plan of action. A common cause of the lapse is an interruption that leads to forgetting the evaluation of the current state of the environment. These lead to mistakes, not slips, because the goals and plans become wrong. Forgetting earlier evaluations often means remaking the decision, sometimes erroneously.

The design cures for memory-lapse mistakes are the same as for memory-lapse slips: ensure that all the relevant information is continuously available. The goals, plans, and current evaluation of the system are of particular importance and should be continually available. Far too many designs eliminate all signs of these items once they have been made or acted upon. Once again, the designer should assume that people will be interrupted during their activities and that they may need assistance in resuming their operations.

Social and Institutional Pressures

A subtle issue that seems to figure in many accidents is social pressure. Although at first it may not seem relevant to design, it has strong influence on everyday behavior. In industrial settings, social pressures can lead to misinterpretation, mistakes, and accidents. To understand human error, it is essential to understand social pressure.

Complex problem-solving is required when one is faced with knowledge-based problems. In some cases, it can take teams of people days to understand what is wrong and the best ways to respond. This is especially true of situations where mistakes have been made in the diagnosis of the problem. Once the mistaken diagnosis is made, all information from then on is interpreted from the wrong point of view. Appropriate reconsiderations might only take place during team turnover, when new people come into the situation with a fresh viewpoint, allowing them to form different interpretations of the events. Sometimes just asking one or more of the team members to take a few hours' break can lead to the same fresh analysis (although it is understandably difficult to convince someone who is battling an emergency situation to stop for a few hours).

In commercial installations, the pressure to keep systems running is immense. Considerable money might be lost if an expensive system is shut down. Operators are often under pressure not to do this. The result has at times been tragic. Nuclear power plants are kept running longer than is safe. Airplanes have taken off before everything was ready and before the pilots had received permission. One such incident led to the largest accident in aviation history. Although the incident happened in 1977, a long time ago, the lessons learned are still very relevant today.

In Tenerife, in the Canary Islands, a KLM Boeing 747 crashed during takeoff into a Pan American 747 that was taxiing on the same runway, killing 583 people. The KLM plane had not received clearance to take off, but the weather was starting to get bad and the crew had already been delayed for too long (even being on the

Canary Islands was a diversion from the scheduled flight—bad weather had prevented their landing at their scheduled destination). And the Pan American flight should not have been on the runway, but there was considerable misunderstanding between the pilots and the air traffic controllers. Furthermore, the fog was coming in so thickly that neither plane's crew could see the other.

In the Tenerife disaster, time and economic pressures were acting together with cultural and weather conditions. The Pan American pilots questioned their orders to taxi on the runway, but they continued anyway. The first officer of the KLM flight voiced minor objections to the captain, trying to explain that they were not yet cleared for takeoff (but the first officer was very junior to the captain, who was one of KLM's most respected pilots). All in all, a major tragedy occurred due to a complex mixture of social pressures and logical explaining away of discrepant observations.

You may have experienced similar pressure, putting off refueling or recharging your car until it was too late and you ran out, sometimes in a truly inconvenient place (this has happened to me). What are the social pressures to cheat on school examinations, or to help others cheat? Or to not report cheating by others? Never underestimate the power of social pressures on behavior, causing otherwise sensible people to do things they know are wrong and possibly dangerous.

When I was in training to do underwater (scuba) diving, our instructor was so concerned about this that he said he would reward anyone who stopped a dive early in favor of safety. People are normally buoyant, so they need weights to get them beneath the surface. When the water is cold, the problem is intensified because divers must then wear either wet or dry suits to keep warm, and these suits add buoyancy. Adjusting buoyancy is an important part of the dive, so along with the weights, divers also wear air vests into which they continually add or remove air so that the body is close to neutral buoyancy. (As divers go deeper, increased water pressure compresses the air in their protective suits and lungs, so they become heavier: the divers need to add air to their vests to compensate.)

When divers have gotten into difficulties and needed to get to the surface quickly, or when they were at the surface close to shore but being tossed around by waves, some drowned because they were still being encumbered by their heavy weights. Because the weights are expensive, the divers didn't want to release them. In addition, if the divers released the weights and then made it back safely, they could never prove that the release of the weights was necessary, so they would feel embarrassed, creating self-induced social pressure. Our instructor was very aware of the resulting reluctance of people to take the critical step of releasing their weights when they weren't entirely positive it was necessary. To counteract this tendency, he announced that if anyone dropped the weights for safety reasons, he would publicly praise the diver and replace the weights at no cost to the person. This was a very persuasive attempt to overcome social pressures.

Social pressures show up continually. They are usually difficult to document because most people and organizations are reluctant to admit these factors, so even if they are discovered in the process of the accident investigation, the results are often kept hidden from public scrutiny. A major exception is in the study of transportation accidents, where the review boards across the world tend to hold open investigations. The US National Transportation Safety Board (NTSB) is an excellent example of this, and its reports are widely used by many accident investigators and researchers of human error (including me).

Another good example of social pressures comes from yet another airplane incident. In 1982 an Air Florida flight from National Airport, Washington, DC, crashed during takeoff into the Fourteenth Street Bridge over the Potomac River, killing seventy-eight people, including four who were on the bridge. The plane should not have taken off because there was ice on the wings, but it had already been delayed for over an hour and a half; this and other factors, the NTSB reported, "may have predisposed the crew to hurry." The accident occurred despite the first officer's attempt to warn the captain, who was flying the airplane (the captain and first officer—sometimes called the copilot—usually alternate flying roles on different legs of a trip). The NTSB report quotes the flight deck recorder's documenting that "although the first officer expressed concern that something 'was not right' to the captain four times during the takeoff, the captain took no action to reject the takeoff." NTSB summarized the causes this way:

> *The National Transportation Safety Board determines that the probable cause of this accident was the flight crew's failure to use engine anti-ice during ground operation and takeoff, their decision to take off with snow/ice on the airfoil surfaces of the aircraft, and the captain's failure to reject the takeoff during the early stage when his attention was called to anomalous engine instrument readings.* (NTSB, 1982.)

Again we see social pressures coupled with time and economic forces.

Social pressures can be overcome, but they are powerful and pervasive. We drive when drowsy or after drinking, knowing full well the dangers, but talking ourselves into believing that we are exempt. How can we overcome these kinds of social problems? Good design alone is not sufficient. We need different training; we need to reward safety and put it above economic pressures. It helps if the equipment can make the potential dangers visible and explicit, but this is not always possible. To adequately address social, economic, and cultural pressures and to improve upon company policies are the hardest parts of ensuring safe operation and behavior.

CHECKLISTS

Checklists are powerful tools, proven to increase the accuracy of behavior and to reduce error, particularly slips and memory lapses. They are especially important in situations with multiple, complex requirements, and even more so where there are interruptions. With multiple people involved in a task, it is essential that the lines of responsibility be clearly spelled out. It is always better to have two people do checklists together as a team: one to read the instruction, the other to execute it. If, instead, a single person executes the checklist and then, later, a second person checks the items, the results are not as robust. The person following the checklist, feeling confident that any errors would be caught, might do the steps too quickly. But the same bias affects the checker. Confident in the ability of the first person, the checker often does a quick, less than thorough job.

One paradox of groups is that quite often, adding more people to check a task makes it less likely that it will be done right. Why? Well, if you were responsible for checking the correct readings on a row of fifty gauges and displays, but you know that two people before you had checked them and that one or two people who come after you will check your work, you might relax, thinking that you don't have to be extra careful. After all, with so many people looking, it would be impossible for a problem to exist without detection. But if everyone thinks the same way, adding more checks can actually increase the chance of error. A collaboratively followed checklist is an effective way to counteract these natural human tendencies.

In commercial aviation, collaboratively followed checklists are widely accepted as essential tools for safety. The checklist is done by two people, usually the two pilots of the airplane (the captain and first officer). In aviation, checklists have proven their worth and are now required in all US commercial flights. But despite the strong evidence confirming their usefulness, many industries still fiercely resist them. It makes people feel that their competence is being questioned. Moreover, when two people are involved, a junior person (in aviation, the first officer) is being asked to watch over the action of the senior person. This is a strong violation of the lines of authority in many cultures.

Physicians and other medical professionals have strongly resisted the use of checklists. It is seen as an insult to their professional competence. "Other people might need checklists," they complain, "but not me." Too bad. Too err is human: we all are subject to slips and mistakes when under stress, or under time or social pressure, or after being subjected to multiple interruptions, each essential in its own right. It is not a threat to professional competence to be human. Legitimate criticisms of particular checklists are used as an indictment against the concept of checklists. Fortunately, checklists are slowly starting to gain acceptance in medical situations. When senior personnel insist on the use of checklists, it actually enhances their authority and professional status. It took decades for checklists to be accepted in commercial aviation: let us hope that medicine and other professions will change more rapidly.

Designing an effective checklist is difficult. The design needs to be iterative, always being refined, ideally using the human-centered design principles of Chapter 6, continually adjusting the list until it covers the essential items yet is not burdensome to perform. Many people who object to checklists are actually objecting to badly designed lists: designing a checklist for a complex task is best done by professional designers in conjunction with subject matter experts.

Printed checklists have one major flaw: they force the steps to follow a sequential ordering, even where this is not necessary or even possible. With complex tasks, the order in which many operations are performed may not matter, as long as they are all completed. Sometimes items early in the list cannot be done at the time they are encountered in the checklist. For example, in aviation one of the steps is to check the amount of fuel in the plane. But what if the fueling operation has not yet been completed when this checklist item is encountered? Pilots will skip over it, intending to come back to it after the plane has been refueled. This is a clear opportunity for a memory-lapse error.

In general, it is bad design to impose a sequential structure to task execution unless the task itself requires it. This is one of the major benefits of electronic checklists: they can keep track of skipped items and can ensure that the list will not be marked as complete until all items have been done.

Reporting Error

If errors can be caught, then many of the problems they might lead to can often be avoided. But not all errors are easy to detect. Moreover, social pressures often make it difficult for people to admit to their own errors (or to report the errors of others). If people report their own errors, they might be fined or punished. Moreover, their friends may make fun of them. If a person reports that someone else made an error, this may lead to severe personal repercussions. Finally, most institutions do not wish to reveal errors made by their staff. Hospitals, courts, police systems, utility companies—all are reluctant to admit to the public that their workers are capable of error. These are all unfortunate attitudes.

The only way to reduce the incidence of errors is to admit their existence, to gather together information about them, and thereby to be able to make the appropriate changes to reduce their occurrence. In the absence of data, it is difficult or impossible to make improvements. Rather than stigmatize those who admit to error, we should thank those who do so and encourage the reporting. We need to make it easier to report errors, for the goal is not to punish, but to determine how it occurred and change things so that it will not happen again.

CASE STUDY: *JIDOKA*—HOW TOYOTA HANDLES ERROR

The Toyota automobile company has developed an extremely efficient error-reduction process for manufacturing, widely known as the Toyota Production System. Among its many key principles is a philosophy called *Jidoka,* which Toyota says is "roughly translated as 'automation with a human touch.'" If a worker notices something wrong, the worker is supposed to report it, sometimes even stopping the entire assembly line if a faulty part is about to proceed to the next station. (A special cord, called an *andon,* stops the assembly line and alerts the expert crew.) Experts converge upon the problem area to determine the cause. "Why did it happen?" "Why was that?" "Why is that the reason?" The philosophy is to ask "Why?" as many times as may be necessary to get to the root cause of the problem and then fix it so it can never occur again.

As you might imagine, this can be rather discomforting for the person who found the error. But the report is expected, and when it is discovered that people have failed to report errors, they are punished, all in an attempt to get the workers to be honest.

POKA-YOKE: ERROR PROOFING

Poka-yoke is another Japanese method, this one invented by Shigeo Shingo, one of the Japanese engineers who played a major role in the development of the Toyota Production System. *Poka-yoke* translates as "error proofing" or "avoiding error." One of the techniques of poka-yoke is to add simple fixtures, jigs, or devices to constrain the operations so that they are correct. I practice this myself in my home. One trivial example is a device to help me remember which way to turn the key on the many doors in the apartment complex where I live. I went around with a pile of small, circular, green stick-on dots and put them on each door beside its keyhole, with the green dot indicating the direction in which the key needed to be turned: I added signifiers to the doors. Is this a major error? No. But eliminating it has proven to be convenient. (Neighbors have commented on their utility, wondering who put them there.)

In manufacturing facilities, poka-yoke might be a piece of wood to help align a part properly, or perhaps plates designed with asymmetrical screw holes so that the plate could fit in only one position. Covering emergency or critical switches with a cover to prevent accidental triggering is another poka-yoke technique: this is obviously a forcing function. All the poka-yoke techniques involve a combination of the principles discussed in this book: affordances, signifiers, mapping, and constraints, and perhaps most important of all, forcing functions.

NASA'S AVIATION SAFETY REPORTING SYSTEM

US commercial aviation has long had an extremely effective system for encouraging pilots to submit reports of errors. The program has resulted in numerous improvements to aviation safety. It wasn't easy to establish: pilots had severe self-induced social pressures against admitting to errors. Moreover, to whom would they report them? Certainly not to their employers. Not even to the Federal Aviation Authority (FAA), for then they would probably be punished. The solution was to let the National Aeronautics and Space Administration (NASA) set up a voluntary accident reporting system whereby pilots could submit semi-anonymous reports

of errors they had made or observed in others (semi-anonymous because pilots put their name and contact information on the reports so that NASA could call to request more information). Once NASA personnel had acquired the necessary information, they would detach the contact information from the report and mail it back to the pilot. This meant that NASA no longer knew who had reported the error, which made it impossible for the airline companies or the FAA (which enforced penalties against errors) to find out who had submitted the report. If the FAA had independently noticed the error and tried to invoke a civil penalty or certificate suspension, the receipt of self-report automatically exempted the pilot from punishment (for minor infractions).

When a sufficient number of similar errors had been collected, NASA would analyze them and issue reports and recommendations to the airlines and to the FAA. These reports also helped the pilots realize that their error reports were valuable tools for increasing safety. As with checklists, we need similar systems in the field of medicine, but it has not been easy to set up. NASA is a neutral body, charged with enhancing aviation safety, but has no oversight authority, which helped gain the trust of pilots. There is no comparable institution in medicine: physicians are afraid that self-reported errors might lead them to lose their license or be subjected to lawsuits. But we can't eliminate errors unless we know what they are. The medical field is starting to make progress, but it is a difficult technical, political, legal, and social problem.

Detecting Error

Errors do not necessarily lead to harm if they are discovered quickly. The different categories of errors have differing ease of discovery. In general, action slips are relatively easy to discover; mistakes, much more difficult. Action slips are relatively easy to detect because it is usually easy to notice a discrepancy between the intended act and the one that got performed. But this detection can only take place if there is feedback. If the result of the action is not visible, how can the error be detected?

Memory-lapse slips are difficult to detect precisely because there is nothing to see. With a memory slip, the required action is not performed. When no action is done, there is nothing to detect. It is only when the lack of action allows some unwanted event to occur that there is hope of detecting a memory-lapse slip.

Mistakes are difficult to detect because there is seldom anything that can signal an inappropriate goal. And once the wrong goal or plan is decided upon, the resulting actions are consistent with that wrong goal, so careful monitoring of the actions not only fails to detect the erroneous goal, but, because the actions are done correctly, can inappropriately provide added confidence to the decision.

Faulty diagnoses of a situation can be surprisingly difficult to detect. You might expect that if the diagnosis was wrong, the actions would turn out to be ineffective, so the fault would be discovered quickly. But misdiagnoses are not random. Usually they are based on considerable knowledge and logic. The misdiagnosis is usually both reasonable and relevant to eliminating the symptoms being observed. As a result, the initial actions are apt to appear appropriate and helpful. This makes the problem of discovery even more difficult. The actual error might not be discovered for hours or days.

Memory-lapse mistakes are especially difficult to detect. Just as with a memory-lapse slip the absence of something that should have been done is always more difficult to detect than the presence of something that should not have been done. The difference between memory-lapse slips and mistakes is that, in the first case, a single component of a plan is skipped, whereas in the second, the entire plan is forgotten. Which is easier to discover? At this point I must retreat to the standard answer science likes to give to questions of this sort: "It all depends."

EXPLAINING AWAY MISTAKES

Mistakes can take a long time to be discovered. Hear a noise that sounds like a pistol shot and think: "Must be a car's exhaust backfiring." Hear someone yell outside and think: "Why can't my neighbors be quiet?" Are we correct in dismissing these incidents? Most of the time we are, but when we're not, our explanations can be difficult to justify.

Explaining away errors is a common problem in commercial accidents. Most major accidents are preceded by warning signs: equipment malfunctions or unusual events. Often, there is a series of apparently unrelated breakdowns and errors that culminate in major disaster. Why didn't anyone notice? Because no single incident appeared to be serious. Often, the people involved noted each problem but discounted it, finding a logical explanation for the otherwise deviant observation.

THE CASE OF THE WRONG TURN ON A HIGHWAY

I've misinterpreted highway signs, as I'm sure most drivers have. My family was traveling from San Diego to Mammoth Lakes, California, a ski area about 400 miles north. As we drove, we noticed more and more signs advertising the hotels and gambling casinos of Las Vegas, Nevada. "Strange," we said, "Las Vegas always did advertise a long way off—there is even a billboard in San Diego—but this seems excessive, advertising on the road to Mammoth." We stopped for gasoline and continued on our journey. Only later, when we tried to find a place to eat supper, did we discover that we had missed a turn nearly two hours earlier, before we had stopped for gasoline, and that we were actually on the road to Las Vegas, not the road to Mammoth. We had to backtrack the entire two-hour segment, wasting four hours of driving. It's humorous now; it wasn't then.

Once people find an explanation for an apparent anomaly, they tend to believe they can now discount it. But explanations are based on analogy with past experiences, experiences that may not apply to the current situation. In the driving story, the prevalence of billboards for Las Vegas was a signal we should have heeded, but it seemed easily explained. Our experience is typical: some major industrial incidents have resulted from false explanations of anomalous events. But do note: usually these apparent anomalies should be ignored. Most of the time, the explanation for their presence is correct. Distinguishing a true anomaly from an apparent one is difficult.

IN HINDSIGHT, EVENTS SEEM LOGICAL

The contrast in our understanding before and after an event can be dramatic. The psychologist Baruch Fischhoff has studied explanations given in hindsight, where events seem completely obvious and predictable after the fact but completely unpredictable beforehand.

Fischhoff presented people with a number of situations and asked them to predict what would happen: they were correct only at the chance level. When the actual outcome was not known by the people being studied, few predicted the actual outcome. He then presented the same situations along with the actual outcomes to another group of people, asking them to state how likely each outcome was: when the actual outcome was known, it appeared to be plausible and likely and other outcomes appeared unlikely.

Hindsight makes events seem obvious and predictable. Foresight is difficult. During an incident, there are never clear clues. Many things are happening at once: workload is high, emotions and stress levels are high. Many things that are happening will turn out to be irrelevant. Things that appear irrelevant will turn out to be critical. The accident investigators, working with hindsight, knowing what really happened, will focus on the relevant information and ignore the irrelevant. But at the time the events were happening, the operators did not have information that allowed them to distinguish one from the other.

This is why the best accident analyses can take a long time to do. The investigators have to imagine themselves in the shoes of the people who were involved and consider all the information, all the training, and what the history of similar past events would have taught the operators. So, the next time a major accident occurs, ignore the initial reports from journalists, politicians, and executives who don't have any substantive information but feel compelled to provide statements anyway. Wait until the official reports come from trusted sources. Unfortunately, this could be months or years after the accident, and the public usually wants answers immediately, even if those answers are wrong. Moreover, when the full story finally appears, newspapers will no longer consider it news, so they won't report it. You will have to search for the official report. In the United States, the National Transportation Safety Board (NTSB) can be trusted. NTSB conducts careful investigations of all major aviation, automobile and truck, train, ship, and pipeline incidents. (Pipelines? Sure: pipelines transport coal, gas, and oil.)

Designing for Error

It is relatively easy to design for the situation where everything goes well, where people use the device in the way that was intended, and no unforeseen events occur. The tricky part is to design for when things go wrong.

Consider a conversation between two people. Are errors made? Yes, but they are not treated as such. If a person says something that is not understandable, we ask for clarification. If a person says something that we believe to be false, we question and debate. We don't issue a warning signal. We don't beep. We don't give error messages. We ask for more information and engage in mutual dialogue to reach an understanding. In normal conversations between two friends, misstatements are taken as normal, as approximations to what was really meant. Grammatical errors, self-corrections, and restarted phrases are ignored. In fact, they are usually not even detected because we concentrate upon the intended meaning, not the surface features.

Machines are not intelligent enough to determine the meaning of our actions, but even so, they are far less intelligent than they could be. With our products, if we do something inappropriate, if the action fits the proper format for a command, the product does it, even if it is outrageously dangerous. This has led to tragic accidents, especially in health care, where inappropriate design of infusion pumps and X-ray machines allowed extreme overdoses of medication or radiation to be administered to patients, leading to their deaths. In financial institutions, simple keyboard errors have led to huge financial transactions, far beyond normal limits.

Even simple checks for reasonableness would have stopped all of these errors. (This is discussed at the end of the chapter under the heading "Sensibility Checks.")

Many systems compound the problem by making it easy to err but difficult or impossible to discover error or to recover from it. It should not be possible for one simple error to cause widespread damage. Here is what should be done:

- Understand the causes of error and design to minimize those causes.
- Do sensibility checks. Does the action pass the "common sense" test?
- Make it possible to reverse actions—to "undo" them—or make it harder to do what cannot be reversed.
- Make it easier for people to discover the errors that do occur, and make them easier to correct.
- Don't treat the action as an error; rather, try to help the person complete the action properly. Think of the action as an approximation to what is desired.

As this chapter demonstrates, we know a lot about errors. Thus, novices are more likely to make mistakes than slips, whereas experts are more likely to make slips. Mistakes often arise from ambiguous or unclear information about the current state of a system, the lack of a good conceptual model, and inappropriate procedures. Recall that most mistakes result from erroneous choice of goal or plan or erroneous evaluation and interpretation. All of these come about through poor information provided by the system about the choice of goals and the means to accomplish them (plans), and poor-quality feedback about what has actually happened.

A major source of error, especially memory-lapse errors, is interruption. When an activity is interrupted by some other event, the cost of the interruption is far greater than the loss of the time required to deal with the interruption: it is also the cost of resuming the interrupted activity. To resume, it is necessary to remember precisely the previous state of the activity: what the goal was, where one was in the action cycle, and the relevant state of the system. Most systems make it difficult to resume after an interruption.

Most discard critical information that is needed by the user to remember the numerous small decisions that had been made, the things that were in the person's short-term memory, to say nothing of the current state of the system. What still needs to be done? Maybe I was finished? It is no wonder that many slips and mistakes are the result of interruptions.

Multitasking, whereby we deliberately do several tasks simultaneously, erroneously appears to be an efficient way of getting a lot done. It is much beloved by teenagers and busy workers, but in fact, all the evidence points to severe degradation of performance, increased errors, and a general lack of both quality and efficiency. Doing two tasks at once takes longer than the sum of the times it would take to do each alone. Even as simple and common a task as talking on a hands-free cell phone while driving leads to serious degradation of driving skills. One study even showed that cell phone usage during walking led to serious deficits: "Cell phone users walked more slowly, changed directions more frequently, and were less likely to acknowledge other people than individuals in the other conditions. In the second study, we found that cell phone users were less likely to notice an unusual activity along their walking route (a unicycling clown)" (Hyman, Boss, Wise, McKenzie, & Caggiano, 2010).

A large percentage of medical errors are due to interruptions. In aviation, where interruptions were also determined to be a major problem during the critical phases of flying—landing and takeoff—the US Federal Aviation Authority (FAA) requires what it calls a "Sterile Cockpit Configuration," whereby pilots are not allowed to discuss any topic not directly related to the control of the airplane during these critical periods. In addition, the flight attendants are not permitted to talk to the pilots during these phases (which has at times led to the opposite error—failure to inform the pilots of emergency situations).

Establishing similar sterile periods would be of great benefit to many professions, including medicine and other safety-critical operations. My wife and I follow this convention in driving: when the driver is entering or leaving a high-speed highway, conversation ceases until the transition has been completed. Interruptions and distractions lead to errors, both mistakes and slips.

Warning signals are usually not the answer. Consider the control room of a nuclear power plant, the cockpit of a commercial aircraft, or the operating room of a hospital. Each has a large number of different instruments, gauges, and controls, all with signals that tend to sound similar because they all use simple tone generators to beep their warnings. There is no coordination among the instruments, which means that in major emergencies, they all sound at once. Most can be ignored anyway because they tell the operator about something that is already known. Each competes with the others to be heard, interfering with efforts to address the problem.

Unnecessary, annoying alarms occur in numerous situations. How do people cope? By disconnecting warning signals, taping over warning lights (or removing the bulbs), silencing bells, and basically getting rid of all the safety warnings. The problem comes after such alarms are disabled, either when people forget to restore the warning systems (there are those memory-lapse slips again), or if a different incident happens while the alarms are disconnected. At that point, nobody notices. Warnings and safety methods must be used with care and intelligence, taking into account the tradeoffs for the people who are affected.

The design of warning signals is surprisingly complex. They have to be loud or bright enough to be noticed, but not so loud or bright that they become annoying distractions. The signal has to both attract attention (act as a signifier of critical information) and also deliver information about the nature of the event that is being signified. The various instruments need to have a coordinated response, which means that there must be international standards and collaboration among the many design teams from different, often competing, companies. Although considerable research has been directed toward this problem, including the development of national standards for alarm management systems, the problem still remains in many situations.

More and more of our machines present information through speech. But like all approaches, this has both strengths and weaknesses. It allows for precise information to be conveyed, especially when the person's visual attention is directed elsewhere. But if several speech warnings operate at the same time, or if the environment is noisy, speech warnings may not be understood. Or if conversations among the users or operators are necessary, speech warnings will interfere. Speech warning signals can be effective, but only if used intelligently.

DESIGN LESSONS FROM THE STUDY OF ERRORS

Several design lessons can be drawn from the study of errors, one for preventing errors before they occur and one for detecting and correcting them when they do occur. In general, the solutions follow directly from the preceding analyses.

ADDING CONSTRAINTS TO BLOCK ERRORS

Prevention often involves adding specific constraints to actions. In the physical world, this can be done through clever use of shape and size. For example, in automobiles, a variety of fluids are required for safe operation and maintenance: engine oil, transmission oil, brake fluid, windshield washer solution, radiator coolant, battery water, and gasoline. Putting the wrong fluid into a reservoir could lead to serious damage or even an accident. Automobile manufacturers try to minimize these errors by segregating the filling points, thereby reducing description-similarity errors. When the filling points for fluids that should be added only occasionally or by qualified mechanics are located separately from those for fluids used more frequently, the average motorist is unlikely to use the incorrect filling points. Errors in adding fluids to the wrong container can be minimized by making the openings have different sizes and shapes, providing physical constraints against inappropriate filling. Different fluids often have different colors so that they can be distinguished. All these are excellent ways to minimize errors. Similar techniques are in widespread use in hospitals and industry. All of these are intelligent applications of constraints, forcing functions, and poka-yoke.

Electronic systems have a wide range of methods that could be used to reduce error. One is to segregate controls, so that easily confused controls are located far from one another. Another is to use separate modules, so that any control not directly relevant to the current operation is not visible on the screen, but requires extra effort to get to.

UNDO

Perhaps the most powerful tool to minimize the impact of errors is the Undo command in modern electronic systems, reversing the operations performed by the previous command, wherever possible. The best systems have multiple levels of undoing, so it is possible to undo an entire sequence of actions.

Obviously, undoing is not always possible. Sometimes, it is only effective if done immediately after the action. Still, it is a powerful tool to minimize the impact of error. It is still amazing to me that many electronic and computer-based systems fail to provide a means to undo even where it is clearly possible and desirable.

CONFIRMATION AND ERROR MESSAGES

Many systems try to prevent errors by requiring confirmation before a command will be executed, especially when the action will destroy something of importance. But these requests are usually ill-timed because after requesting an operation, people are usually certain they want it done. Hence the standard joke about such warnings:

Person: Delete "my most important file."
System: Do you want to delete "my most important file"?
Person: Yes.
System: Are you certain?
Person: Yes!
System "My most favorite file" has been deleted.
Person: Oh. Damn.

The request for confirmation seems like an irritant rather than an essential safety check because the person tends to focus upon the action rather than the object that is being acted upon. A better check would be a prominent display of both the action to be taken and the object, perhaps with the choice of "cancel" or "do it." The important point is making salient what the implications of the action are. Of course, it is because of errors of this sort that the Undo command is so important. With traditional graphical user interfaces on computers, not only is Undo a standard command, but when files are "deleted," they are actually simply moved from sight and stored in the file folder named "Trash," so that in the above example, the person could open the Trash and retrieve the erroneously deleted file.

Confirmations have different implications for slips and mistakes. When I am writing, I use two very large displays and a powerful computer. I might have seven to ten applications running simultaneously. I have sometimes had as many as forty open windows. Suppose I activate the command that closes one of the windows, which triggers a confirmatory message: did I wish to close the window? How I deal with this depends upon why I requested that the window be closed. If it was a slip, the confirmation required will be useful. If it was by mistake, I am apt to ignore it. Consider these two examples:

A slip leads me to close the wrong window.

Suppose I intended to type the word *We*, but instead of typing Shift + W for the first character, I typed Command + W (or Control + W), the keyboard command for closing a window. Because I expected the screen to display an uppercase W, when a dialog box appeared, asking whether I really wanted to delete the file, I would be surprised, which would immediately alert me to the slip. I would cancel the action (an alternative thoughtfully provided by the dialog box) and retype the Shift + W, carefully this time.

A mistake leads me to close the wrong window.

Now suppose I really intended to close a window. I often use a temporary file in a window to keep notes about the chapter I am working on. When I am finished with it, I close it without saving its contents—after all, I am finished. But because I usually have multiple windows open, it is very easy to close the wrong one. The computer assumes that all commands apply to the active window—the one where the last actions had been performed (and which contains the text cursor). But if I reviewed the temporary window prior to closing it, my visual attention is focused upon that window, and when I decide to close it, I forget that it is not the active window from the computer's point of view. So I issue the command to shut the window, the computer presents me with a dialog box, asking for confirmation, and I accept it, choosing the option not to save my work. Because the dialog box was expected, I didn't bother to read it. As a result, I closed the wrong window and worse, did not save any of the typing, possibly losing considerable work. Warning messages are surprisingly ineffective against mistakes (even nice requests, such as the one shown in Chapter 4, Figure 4.6, page 143).

Was this a mistake or a slip? Both. Issuing the "close" command while the wrong window was active is a memory-lapse slip. But deciding not to read the dialog box and accepting it without saving the contents is a mistake (two mistakes, actually).

What can a designer do? Several things:

- **Make the item being acted upon more prominent.** That is, change the appearance of the actual object being acted upon to be more visible: enlarge it, or perhaps change its color.
- **Make the operation reversible.** If the person saves the content, no harm is done except the annoyance of having to reopen the file. If the person elects Don't Save, the system could secretly save the contents, and the next time the person opened the file, it could ask whether it should restore it to the latest condition.

SENSIBILITY CHECKS

Electronic systems have another advantage over mechanical ones: they can check to make sure that the requested operation is sensible.

It is amazing that in today's world, medical personnel can accidentally request a radiation dose a thousand times larger than normal and have the equipment meekly comply. In some cases, it isn't even possible for the operator to notice the error.

Similarly, errors in stating monetary sums can lead to disastrous results, even though a quick glance at the amount would indicate that something was badly off. For example, there are roughly 1,000 Korean won to the US dollar. Suppose I wanted to transfer $1,000 into a Korean bank account in *won* ($1,000 is roughly ₩1,000,000). But suppose I enter the Korean number into the dollar field. Oops—I'm trying to transfer a million dollars. Intelligent systems would take note of the normal size of my transactions, querying if the amount was considerably larger than normal. For me, it would query the million-dollar request. Less intelligent systems would blindly follow instructions, even though I did not have a million dollars in my account (in fact, I would probably be charged a fee for overdrawing my account).

Sensibility checks, of course, are also the answer to the serious errors caused when inappropriate values are entered into hospital medication and X-ray systems or in financial transactions, as discussed earlier in this chapter.

MINIMIZING SLIPS

Slips most frequently occur when the conscious mind is distracted, either by some other event or simply because the action being performed is so well learned that it can be done automatically, without conscious attention. As a result, the person does not pay sufficient attention to the action or its consequences. It might therefore seem that one way to minimize slips is to ensure that people always pay close, conscious attention to the acts being done.

Bad idea. Skilled behavior is subconscious, which means it is fast, effortless, and usually accurate. Because it is so automatic, we can type at high speeds even while the conscious mind is occupied composing the words. This is why we can walk and talk while navigating traffic and obstacles. If we had to pay conscious attention to every little thing we did, we would accomplish far less in our lives. The information processing structures of the brain automatically regulate how much conscious attention is being paid to a task: conversations automatically pause when crossing the street amid busy traffic. Don't count on it, though: if too much attention is focused on something else, the fact that the traffic is getting dangerous might not be noted.

Many slips can be minimized by ensuring that the actions and their controls are as dissimilar as possible, or at least, as physically far apart as possible. Mode errors can be eliminated by the simple expedient of eliminating most modes and, if this is not possible, by making the modes very visible and distinct from one another.

The best way of mitigating slips is to provide perceptible feedback about the nature of the action being performed, then very perceptible feedback describing the new resulting state, coupled with a mechanism that allows the error to be undone. For example, the use of machine-readable codes has led to a dramatic reduction in the delivery of wrong medications to patients. Prescriptions sent to the pharmacy are given electronic codes, so the pharmacist can scan both the prescription and the resulting medication to ensure they are the same. Then, the nursing staff at the hospital scans both the label of the medication and the tag worn around the patient's wrist to ensure that the medication is being given to the correct individual. Moreover, the computer system can flag repeated administration of the same medication. These scans do increase the workload, but only slightly. Other kinds of errors are still possible, but these simple steps have already been proven worthwhile.

Common engineering and design practices seem as if they are deliberately intended to cause slips. Rows of identical controls or meters is a sure recipe for description-similarity errors. Internal modes that are not very conspicuously marked are a clear driver of mode errors. Situations with numerous interruptions, yet where the design assumes undivided attention, are a clear enabler of memory lapses—and almost no equipment today is designed to support the numerous interruptions that so many situations entail. And failure to provide assistance and visible reminders for performing infrequent procedures that are similar to much more frequent ones leads to capture errors, where the more frequent actions are performed rather than the correct ones for the situation. Procedures should be designed so that the initial steps are as dissimilar as possible.

The important message is that good design can prevent slips and mistakes. Design can save lives.

THE SWISS CHEESE MODEL OF HOW ERRORS LEAD TO ACCIDENTS

Fortunately, most errors do not lead to accidents. Accidents often have numerous contributing causes, no single one of which is the root cause of the incident.

James Reason likes to explain this by invoking the metaphor of multiple slices of Swiss cheese, the cheese famous for being riddled with holes (Figure 5.3). If each slice of cheese represents a condition in the task being done, an accident can happen only if holes in all four slices of cheese are lined up just right. In well-designed systems, there can be many equipment failures, many errors, but they will not lead to an accident unless they all line up precisely. Any leakage—passageway through a hole—is most likely blocked at the next level. Well-designed systems are resilient against failure.

This is why the attempt to find "the" cause of an accident is usually doomed to fail. Accident investigators, the press, government officials, and the everyday citizen like to find simple explanations for the cause of an accident. "See, if the hole in slice A

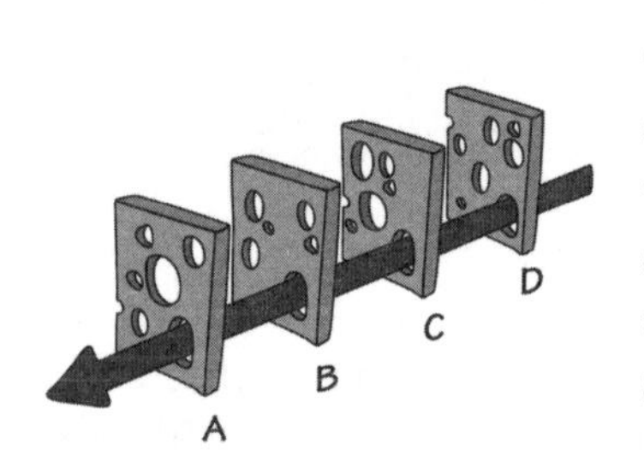

FIGURE 5.3. Reason's Swiss Cheese Model of Accidents. Accidents usually have multiple causes, whereby had any single one of those causes not happened, the accident would not have occurred. The British accident researcher James Reason describes this through the metaphor of slices of Swiss cheese: unless the holes all line up perfectly, there will be no accident. This metaphor provides two lessons: First, do not try to find "the" cause of an accident; Second, we can decrease accidents and make systems more resilient by designing them to have extra precautions against error (more slices of cheese), less opportunities for slips, mistakes, or equipment failure (less holes), and very different mechanisms in the different subparts of the system (trying to ensure that the holes do not line up). (Drawing based upon one by Reason, 1990.)

had been slightly higher, we would not have had the accident. So throw away slice A and replace it." Of course, the same can be said for slices B, C, and D (and in real accidents, the number of cheese slices would sometimes measure in the tens or hundreds). It is relatively easy to find some action or decision that, had it been different, would have prevented the accident. But that does not mean that this was the cause of the accident. It is only one of the many causes: all the items have to line up.

You can see this in most accidents by the "if only" statements. "If only I hadn't decided to take a shortcut, I wouldn't have had the accident." "If only it hadn't been raining, my brakes would have worked." "If only I had looked to the left, I would have seen the car sooner." Yes, all those statements are true, but none of them is "the" cause of the accident. Usually, there is no single cause. Yes, journalists and lawyers, as well as the public, like to know the cause so someone can be blamed and punished. But reputable investigating agencies know that there is not a single cause, which is why their investigations take so long. Their responsibility is to understand the system and make changes that would reduce the chance of the same sequence of events leading to a future accident.

The Swiss cheese metaphor suggests several ways to reduce accidents:

- Add more slices of cheese.
- Reduce the number of holes (or make the existing holes smaller).
- Alert the human operators when several holes have lined up.

Each of these has operational implications. More slices of cheese means mores lines of defense, such as the requirement in aviation and other industries for checklists, where one person reads the items, another does the operation, and the first person checks the operation to confirm it was done appropriately.

Reducing the number of critical safety points where error can occur is like reducing the number or size of the holes in the Swiss cheese. Properly designed equipment will reduce the opportunity for slips and mistakes, which is like reducing the number of holes and making the ones that remain smaller. This is precisely how the safety level of commercial aviation has been dramatically improved. Deborah Hersman, chair of the National Transportation Safety Board, described the design philosophy as:

> *U.S. airlines carry about two million people through the skies safely every day, which has been achieved in large part through design redundancy and layers of defense.*

Design redundancy and layers of defense: that's Swiss cheese. The metaphor illustrates the futility of trying to find the one underlying cause of an accident (usually some person) and punishing the culprit. Instead, we need to think about systems, about all the interacting factors that lead to human error and then to accidents, and devise ways to make the systems, as a whole, more reliable.

When Good Design Isn't Enough

WHEN PEOPLE REALLY ARE AT FAULT

I am sometimes asked whether it is really right to say that people are never at fault, that it is always bad design. That's a sensible question. And yes, of course, sometimes it is the person who is at fault.

Even competent people can lose competency if sleep deprived, fatigued, or under the influence of drugs. This is why we have laws banning pilots from flying if they have been drinking within some specified period and why we limit the number of hours they can fly without rest. Most professions that involve the risk of death or injury have similar regulations about drinking, sleep, and drugs. But everyday jobs do not have these restrictions. Hospitals often require their staff to go without sleep for durations that far exceed the safety requirements of airlines. Why? Would you be happy having a sleep-deprived physician operating on you? Why is sleep deprivation considered dangerous in one situation and ignored in another?

Some activities have height, age, or strength requirements. Others require considerable skills or technical knowledge: people not trained or not competent should not be doing them. That is why many activities require government-approved training and licensing. Some examples are automobile driving, airplane piloting, and medical practice. All require instructional courses and tests. In aviation, it isn't sufficient to be trained: pilots must also keep in practice by flying some minimum number of hours per month.

Drunk driving is still a major cause of automobile accidents: this is clearly the fault of the drinker. Lack of sleep is another major culprit in vehicle accidents. But because people occasionally are at fault does not justify the attitude that assumes they are always at fault. The far greater percentage of accidents is the result of poor design, either of equipment or, as is often the case in industrial accidents, of the procedures to be followed.

As noted in the discussion of deliberate violations earlier in this chapter (page 169), people will sometimes deliberately violate procedures and rules, perhaps because they cannot get their jobs done otherwise, perhaps because they believe there are extenuating circumstances, and sometimes because they are taking the gamble that the relatively low probability of failure does not apply to them. Unfortunately, if someone does a dangerous activity that only results in injury or death one time in a million, that can lead to hundreds of deaths annually across the world, with its 7 billion people. One of my favorite examples in aviation is of a pilot who, after experiencing low oil-pressure readings in all three of his engines, stated that it must be an instrument failure because it was a one-in-a-million chance that the readings were true. He was right in his assessment, but unfortunately, he was the one. In the United States alone there were roughly 9 million flights in 2012. So, a one-in-a-million chance could translate into nine incidents.

Sometimes, people really are at fault.

Resilience Engineering

In industrial applications, accidents in large, complex systems such as oil wells, oil refineries, chemical processing plants, electrical power systems, transportation, and medical services can have major impacts on the company and the surrounding community.

Sometimes the problems do not arise in the organization but outside it, such as when fierce storms, earthquakes, or tidal waves demolish large parts of the existing infrastructure. In either case, the question is how to design and manage these systems so that they can restore services with a minimum of disruption and damage. An important approach is *resilience engineering,* with the goal of designing systems, procedures, management, and the training of people so they are able to respond to problems as they arise. It strives to ensure that the design of all these things—the equipment, procedures, and communication both among workers and also externally to management and the public—are continually being assessed, tested, and improved.

Thus, major computer providers can deliberately cause errors in their systems to test how well the company can respond. This is done by deliberately shutting down critical facilities to ensure that the backup systems and redundancies actually work. Although it might seem dangerous to do this while the systems are online, serving real customers, the only way to test these large, complex systems is by doing so. Small tests and simulations do not carry the complexity, stress levels, and unexpected events that characterize real system failures.

As Erik Hollnagel, David Woods, and Nancy Leveson, the authors of an early influential series of books on the topic, have skillfully summarized:

> *Resilience engineering is a paradigm for safety management that focuses on how to help people cope with complexity under pressure to achieve success. It strongly contrasts with what is typical today—a paradigm of tabulating error as if it were a thing, followed by interventions to reduce this count. A resilient organisation treats safety as a core value, not a commodity that can be counted. Indeed, safety shows itself only by the events that do not happen! Rather than view past success as a reason to ramp down investments, such organisations continue to invest in anticipating the changing potential for failure because they appreciate that their knowledge of the gaps is imperfect and that their environment constantly changes. One measure of resilience is therefore the ability to create foresight—to anticipate the changing shape of risk,*

before failure and harm occurs. (Reprinted by permission of the publishers. Hollnagel, Woods, & Leveson, 2006, p. 6.)

The Paradox of Automation

Machines are getting smarter. More and more tasks are becoming fully automated. As this happens, there is a tendency to believe that many of the difficulties involved with human control will go away. Across the world, automobile accidents kill and injure tens of millions of people every year. When we finally have widespread adoption of self-driving cars, the accident and casualty rate will probably be dramatically reduced, just as automation in factories and aviation have increased efficiency while lowering both error and the rate of injury.

When automation works, it is wonderful, but when it fails, the resulting impact is usually unexpected and, as a result, dangerous. Today, automation and networked electrical generation systems have dramatically reduced the amount of time that electrical power is not available to homes and businesses. But when the electrical power grid goes down, it can affect huge sections of a country and take many days to recover. With self-driving cars, I predict that we will have fewer accidents and injuries, but that when there is an accident, it will be huge.

Automation keeps getting more and more capable. Automatic systems can take over tasks that used to be done by people, whether it is maintaining the proper temperature, automatically keeping an automobile within its assigned lane at the correct distance from the car in front, enabling airplanes to fly by themselves from takeoff to landing, or allowing ships to navigate by themselves. When the automation works, the tasks are usually done as well as or better than by people. Moreover, it saves people from the dull, dreary routine tasks, allowing more useful, productive use of time, reducing fatigue and error. But when the task gets too complex, automation tends to give up. This, of course, is precisely when it is needed the most. The paradox is that automation can take over the dull, dreary tasks, but fail with the complex ones.

When automation fails, it often does so without warning. This is a situation I have documented very thoroughly in my other books and many of my papers, as have many other people in the field of safety and automation. When the failure occurs, the human is "out of the loop." This means that the person has not been paying much attention to the operation, and it takes time for the failure to be noticed and evaluated, and then to decide how to respond.

In an airplane, when the automation fails, there is usually considerable time for the pilots to understand the situation and respond. Airplanes fly quite high: over 10 km (6 miles) above the earth, so even if the plane were to start falling, the pilots might have several minutes to respond. Moreover, pilots are extremely well trained. When automation fails in an automobile, the person might have only a fraction of a second to avoid an accident. This would be extremely difficult even for the most expert driver, and most drivers are not well trained.

In other circumstances, such as ships, there may be more time to respond, but only if the failure of the automation is noticed. In one dramatic case, the grounding of the cruise ship *Royal Majesty* in 1997, the failure lasted for several days and was only detected in the postaccident investigation, after the ship had run aground, causing several million dollars in damage. What happened? The ship's location was normally determined by the Global Positioning System (GPS), but the cable that connected the satellite antenna to the navigation system somehow had become disconnected (nobody ever discovered how). As a result, the navigation system had switched from using GPS signals to "dead reckoning," approximating the ship's location by estimating speed and direction of travel, but the design of the navigation system didn't make this apparent. As a result, as the ship traveled from Bermuda to its destination of Boston, it went too far south and went aground on Cape Cod, a peninsula jutting out of the water south of Boston. The automation had performed flawlessly for years, which increased people's trust and reliance upon it, so the normal manual checking of location or careful perusal of the display (to see the tiny letters "dr" indicating "dead reckoning" mode) were not done. This was a huge mode error failure.

Design Principles for Dealing with Error

People are flexible, versatile, and creative. Machines are rigid, precise, and relatively fixed in their operations. There is a mismatch between the two, one that can lead to enhanced capability if used properly. Think of an electronic calculator. It doesn't do mathematics like a person, but can solve problems people can't. Moreover, calculators do not make errors. So the human plus calculator is a perfect collaboration: we humans figure out what the important problems are and how to state them. Then we use calculators to compute the solutions.

Difficulties arise when we do not think of people and machines as collaborative systems, but assign whatever tasks can be automated to the machines and leave the rest to people. This ends up requiring people to behave in machine like fashion, in ways that differ from human capabilities. We expect people to monitor machines, which means keeping alert for long periods, something we are bad at. We require people to do repeated operations with the extreme precision and accuracy required by machines, again something we are not good at. When we divide up the machine and human components of a task in this way, we fail to take advantage of human strengths and capabilities but instead rely upon areas where we are genetically, biologically unsuited. Yet, when people fail, they are blamed.

What we call "human error" is often simply a human action that is inappropriate for the needs of technology. As a result, it flags a deficit in our technology. It should not be thought of as error. We should eliminate the concept of error: instead, we should realize that people can use assistance in translating their goals and plans into the appropriate form for technology.

Given the mismatch between human competencies and technological requirements, errors are inevitable. Therefore, the best designs take that fact as given and seek to minimize the opportunities for errors while also mitigating the consequences. Assume that every possible mishap will happen, so protect against them. Make actions reversible; make errors less costly. Here are key design principles:

- Put the knowledge required to operate the technology in the world. Don't require that all the knowledge must be in the head. Allow for efficient operation when people have learned all the requirements, when they are experts who can perform without the knowledge in the world, but make it possible for non-experts to use the knowledge in the world. This will also help experts who need to perform a rare, infrequently performed operation or return to the technology after a prolonged absence.
- Use the power of natural and artificial constraints: physical, logical, semantic, and cultural. Exploit the power of forcing functions and natural mappings.
- Bridge the two gulfs, the Gulf of Execution and the Gulf of Evaluation. Make things visible, both for execution and evaluation. On the execution side, provide feedforward information: make the options readily available. On the evaluation side, provide feedback: make the results of each action apparent. Make it possible to determine the system's status readily, easily, accurately, and in a form consistent with the person's goals, plans, and expectations.

We should deal with error by embracing it, by seeking to understand the causes and ensuring they do not happen again. We need to assist rather than punish or scold.

DESIGN THINKING

One of my rules in consulting is simple: never solve the problem I am asked to solve. Why such a counterintuitive rule? Because, invariably, the problem I am asked to solve is not the real, fundamental, root problem. It is usually a symptom. Just as in Chapter 5, where the solution to accidents and errors was to determine the real, underlying cause of the events, in design, the secret to success is to understand what the real problem is.

It is amazing how often people solve the problem before them without bothering to question it. In my classes of graduate students in both engineering and business, I like to give them a problem to solve on the first day of class and then listen the next week to their wonderful solutions. They have masterful analyses, drawings, and illustrations. The MBA students show spreadsheets in which they have analyzed the demographics of the potential customer base. They show lots of numbers: costs, sales, margins, and profits. The engineers show detailed drawings and specifications. It is all well done, brilliantly presented.

When all the presentations are over, I congratulate them, but ask: "How do you know you solved the correct problem?" They are puzzled. Engineers and business people are trained to solve problems. Why would anyone ever give them the wrong problem? "Where do you think the problems come from?" I ask. The real world is not like the university. In the university, professors make up artificial problems. In the real world, the problems do not come in nice, neat packages. They have to be discovered. It is all too easy to see only the surface problems and never dig deeper to address the real issues.

Solving the Correct Problem

Engineers and businesspeople are trained to solve problems. Designers are trained to discover the real problems. A brilliant solution to the wrong problem can be worse than no solution at all: solve the correct problem.

Good designers never start by trying to solve the problem given to them: they start by trying to understand what the real issues are. As a result, rather than converge upon a solution, they diverge, studying people and what they are trying to accomplish, generating idea after idea after idea. It drives managers crazy. Managers want to see progress: designers seem to be going backward when they are given a precise problem and instead of getting to work, they ignore it and generate new issues to consider, new directions to explore. And not just one, but many. What is going on?

The key emphasis of this book is the importance of developing products that fit the needs and capabilities of people. Design can be driven by many different concerns. Sometimes it is driven by technology, sometimes by competitive pressures or by aesthetics. Some designs explore the limits of technological possibilities; some explore the range of imagination, of society, of art or fashion. Engineering design tends to emphasize reliability, cost, and efficiency. The focus of this book, and of the discipline called human-centered design, is to ensure that the result fits human desires, needs, and capabilities. After all, why do we make products? We make them for people to use.

Designers have developed a number of techniques to avoid being captured by too facile a solution. They take the original problem as a suggestion, not as a final statement, then think broadly about what the issues underlying this problem statement might really be (as was done through the "Five Whys" approach to getting at the root cause, described in Chapter 5). Most important of all is that the process be iterative and expansive. Designers resist the temptation to jump immediately to a solution for the stated problem. Instead, they first spend time determining what basic, fundamental (root) issue needs to be addressed. They don't try to search for a solution until they have determined the real problem, and even then, instead of solving that problem, they stop to consider a wide range of potential solutions. Only then will they finally converge upon their proposal. This process is called *design thinking.*

Design thinking is not an exclusive property of designers—all great innovators have practiced this, even if unknowingly, regardless of whether they were artists or poets, writers or scientists, engineers or businesspeople. But because designers pride themselves on their ability to innovate, to find creative solutions to fundamental problems, design thinking has become the hallmark of the modern design firm. Two of the powerful tools of design thinking are human-centered design and the double-diamond diverge-converge model of design.

Human-centered design (HCD) is the process of ensuring that people's needs are met, that the resulting product is understandable and usable, that it accomplishes the desired tasks, and that the experience of use is positive and enjoyable. Effective design needs to satisfy a large number of constraints and concerns, including shape and form, cost and efficiency, reliability and effectiveness, understandability and usability, the pleasure of the appearance, the pride of ownership, and the joy of actual use. HCD is a procedure for addressing these requirements, but with an emphasis on two things: solving the right problem, and doing so in a way that meets human needs and capabilities.

Over time, the many different people and industries that have been involved in design have settled upon a common set of methods for doing HCD. Everyone has his or her own favorite method, but all are variants on the common theme: iterate through the four stages of observation, generation, prototyping, and testing. But even before this, there is one overriding principle: solve the right problem.

These two components of design—finding the right problem and meeting human needs and capabilities—give rise to two phases of the design process. The first phase is to find the right problem, the second is to find the right solution. Both phases use the HCD process. This double-phase approach to design led the British Design Council to describe it as a "double diamond." So that is where we start the story.

The Double-Diamond Model of Design

Designers often start by questioning the problem given to them: they expand the scope of the problem, diverging to examine all the fundamental issues that underlie it. Then they converge upon a single problem statement. During the solution phase of their studies, they first expand the space of possible solutions, the divergence phase. Finally, they converge upon a proposed solution (Figure 6.1). This double diverge-converge pattern was first introduced in 2005 by the British Design Council, which called it the *double-diamond design process model*. The Design Council divided the design process into four stages: "discover" and "define"—for the divergence and convergence phases of finding the right problem,

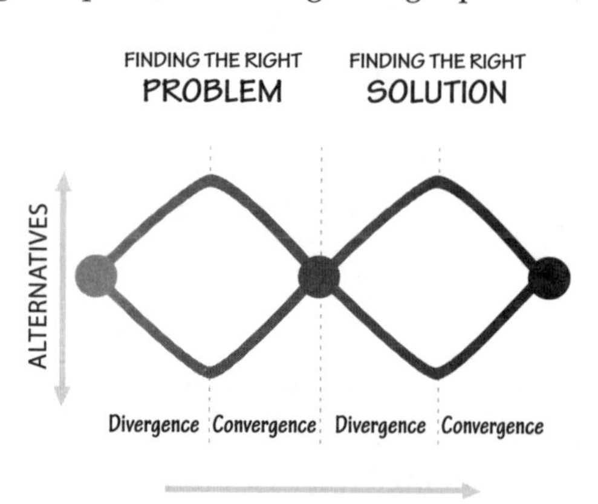

FIGURE 6.1. The Double-Diamond Model of Design. Start with an idea, and through the initial design research, expand the thinking to explore the fundamental issues. Only then is it time to converge upon the real, underlying problem. Similarly, use design research tools to explore a wide variety of solutions before converging upon one. (Slightly modified from the work of the British Design Council, 2005.)

and "develop" and "deliver"—for the divergence and convergence phases of finding the right solution.

The double diverge-converge process is quite effective at freeing designers from unnecessary restrictions to the problem and solution spaces. But you can sympathize with a product manager who, having given the designers a problem to solve, finds them questioning the assignment and insisting on traveling all over the world to seek deeper understanding. Even when the designers start focusing upon the problem, they do not seem to make progress, but instead develop a wide variety of ideas and thoughts, many only half-formed, many clearly impractical. All this can be rather unsettling to the product manager who, concerned about meeting the schedule, wants to see immediate convergence. To add to the frustration of the product manager, as the designers start to converge upon a solution, they may realize that they have inappropriately formulated the problem, so the entire process must be repeated (although it can go more quickly this time).

This repeated divergence and convergence is important in properly determining the right problem to be solved and then the best way to solve it. It looks chaotic and ill-structured, but it actually follows well-established principles and procedures. How does the product manager keep the entire team on schedule despite the apparent random and divergent methods of designers? Encourage their free exploration, but hold them to the schedule (and budget) constraints. There is nothing like a firm deadline to get creative minds to reach convergence.

The Human-Centered Design Process

The double-diamond describes the two phases of design: finding the right problem and fulfilling human needs. But how are these actually done? This is where the human-centered design process comes into play: it takes place within the double-diamond diverge-converge process.

There are four different activities in the human-centered design process (Figure 6.2):

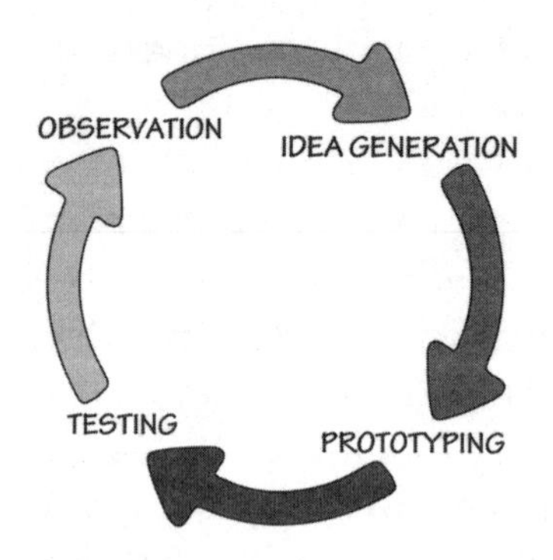

FIGURE 6.2. The Iterative Cycle of Human-Centered Design. Make observations on the intended target population, generate ideas, produce prototypes and test them. Repeat until satisfied. This is often called the *spiral method* (rather than the circle depicted here), to emphasize that each iteration through the stages makes progress.

1. Observation
2. Idea generation (ideation)
3. Prototyping
4. Testing

These four activities are iterated; that is, they are repeated over and over, with each cycle yielding more insights and getting closer to the desired solution. Now let us examine each activity separately.

OBSERVATION

The initial research to understand the nature of the problem itself is part of the discipline of design research. Note that this is research about the customer and the people who will use the products under consideration. It is not the kind of research that scientists do in their laboratories, trying to find new laws of nature. The design researcher will go to the potential customers, observing their activities, attempting to understand their interests, motives, and true needs. The problem definition for the product design will come from this deep understanding of the goals the people are trying to accomplish and the impediments they experience. One of its most critical techniques is to observe the would-be customers in their natural environment, in their normal lives, wherever the product or service being designed will actually be used. Watch them in their homes, schools, and offices. Watch them commute, at parties, at mealtime, and with friends at the local bar. Follow them into the shower if necessary, because it is essential to understand the real situations that they encounter, not some pure isolated experience. This technique is called *applied ethnography,* a method adapted from the field of anthropology. Applied ethnography differs from the slower, more methodical, research-oriented practice of academic anthropologists because the goals are different.

For one, design researchers have the goal of determining human needs that can be addressed through new products. For another, product cycles are driven by schedule and budget, both of which require more rapid assessment than is typical in academic studies that might go on for years.

It's important that the people being observed match those of the intended audience. Note that traditional measures of people, such as age, education, and income, are not always important: what matters most are the activities to be performed. Even when we look at widely different cultures, the activities are often surprisingly similar. As a result, the studies can focus upon the activities and how they get done, while being sensitive to how the local environment and culture might modify those activities. In some cases, such as the products widely used in business, the activity dominates. Thus, automobiles, computers, and phones are pretty standardized across the world because their designs reflect the activities being supported.

In some cases, detailed analyses of the intended group are necessary. Japanese teenage girls are quite different from Japanese women, and in turn, very different from German teenage girls. If a product is intended for subcultures like these, the exact population must be studied. Another way of putting it is that different products serve different needs. Some products are also symbols of status or group membership. Here, although they perform useful functions, they are also fashion statements. This is where teenagers in one culture differ from those of another, and even from younger children and older adults of the same culture. Design researchers must carefully adjust the focus of their observations to the intended market and people for whom the product is intended.

Will the product be used in some country other than where it is being designed? There is only one way to find out: go there (and always include natives in the team). Don't take a shortcut and stay home, talking to students or visitors from that country while remaining in your own: what you will learn is seldom an accurate reflection of the target population or of the ways in which the proposed product will actually be used. There is no substitute for direct observation of and interaction with the people who will be using the product.

Design research supports both diamonds of the design process. The first diamond, finding the right problem, requires a deep understanding of the true needs of people. Once the problem has been defined, finding an appropriate solution again requires deep understanding of the intended population, how those people perform their activities, their capabilities and prior experience, and what cultural issues might be impacted.

DESIGN RESEARCH VERSUS MARKET RESEARCH

Design and marketing are two important parts of the product development group. The two fields are complementary, but each has a different focus. Design wants to know what people really need and how they actually will use the product or service under consideration. Marketing wants to know what people will buy, which includes learning how they make their purchasing decisions. These different aims lead the two groups to develop different methods of inquiry. Designers tend to use qualitative observational methods by which they can study people in depth, understanding how they do their activities and the environmental factors that come into play. These methods are very time consuming, so designers typically only examine small numbers of people, often numbering in the tens.

Marketing is concerned with customers. Who might possibly purchase the item? What factors might entice them to consider and purchase a product? Marketing traditionally uses large-scale, quantitative studies, with heavy reliance on focus groups, surveys, and questionnaires. In marketing, it is not uncommon to converse with hundreds of people in focus groups, and to question tens of thousands of people by means of questionnaires and surveys.

The advent of the Internet and the ability to assess huge amounts of data have given rise to new methods of formal, quantitative market analysis. "Big data," it is called, or sometimes "market analytics." For popular websites, A/B testing is possible in which two potential variants of an offering are tested by giving some randomly selected fraction of visitors (perhaps 10 percent) one set of web pages (the A set); and another randomly selected set of visitors, the other alternative (the B set). In a few hours, hundreds of thousands of visitors may have been exposed to each test set, making it easy to see which yields better results. Moreover, the website can capture a wealth of information about people and their behavior: age, income, home and work addresses, previous purchases, and other websites visited. The virtues of the use of big data for market research are frequently touted. The deficiencies are seldom noted, except for concerns about invasions of personal privacy. In addition to privacy issues, the real problem is that numerical correlations say nothing of people's real needs, of their desires, and of the reasons for their activities. As a result, these numerical data can give a false impression of people. But the use of big data and market analytics is seductive: no travel, little expense, and huge numbers, sexy charts, and impressive statistics, all very persuasive to the executive team trying to decide which new products to develop. After all, what would you trust—neatly presented, colorful charts, statistics, and significance levels based on millions of observations, or the subjective impressions of a motley crew of design researchers who worked, slept, and ate in remote villages, with minimal sanitary facilities and poor infrastructure?

The different methods have different goals and produce very different results. Designers complain that the methods used by marketing don't get at real behavior: what people say they do and want does not correspond with their actual behavior or desires. People in marketing complain that although design research methods yield deep insights, the small number of people observed is a concern. Designers counter with the observation that traditional marketing methods provide shallow insight into a large number of people.

The debate is not useful. All groups are necessary. Customer research is a tradeoff: deep insights on real needs from a tiny set of people, versus broad, reliable purchasing data from a wide range and large number of people. We need both. Designers understand what people really need. Marketing understands what people actually buy. These are not the same things, which is why both approaches are required: marketing and design researchers should work together in complementary teams.

What are the requirements for a successful product? First, if nobody buys the product, then all else is irrelevant. The product design has to provide support for all the factors people use in making purchase decisions. Second, once the product has been purchased and is put into use, it must support real needs so that people can use, understand, and take pleasure from it. The design specifications must include both factors: marketing and design, buying and using.

IDEA GENERATION

Once the design requirements are determined, the next step for a design team is to generate potential solutions. This process is called *idea generation*, or *ideation*. This exercise might be done for both of the double diamonds: during the phase of finding the correct problem, then during the problem solution phase.

This is the fun part of design: it is where creativity is critical. There are many ways of generating ideas: many of these methods fall under the heading of "brainstorming." Whatever the method used, two major rules are usually followed:

- **Generate numerous ideas.** It is dangerous to become fixated upon one or two ideas too early in the process.
- **Be creative without regard for constraints.** Avoid criticizing ideas, whether your own or those of others. Even crazy ideas, often obviously wrong, can contain creative insights that can later be extracted and put to good use in the final idea selection. Avoid premature dismissal of ideas.

I like to add a third rule:

- **Question everything.** I am particularly fond of "stupid" questions. A stupid question asks about things so fundamental that everyone assumes the answer is obvious. But when the question is taken seriously, it often turns out to be profound: the obvious often is not obvious at all. What we assume to be obvious is simply the way things have always been done, but now that it is questioned, we don't actually know the reasons. Quite often the solution to problems is discovered through stupid questions, through questioning the obvious.

PROTOTYPING

The only way to really know whether an idea is reasonable is to test it. Build a quick prototype or mock-up of each potential solution. In the early stages of this process, the mock-ups can be pencil sketches, foam and cardboard models, or simple images made with simple drawing tools. I have made mock-ups with spreadsheets, PowerPoint slides, and with sketches on index cards or sticky notes. Sometimes ideas are best conveyed by skits, especially if you're developing services or automated systems that are difficult to prototype.

One popular prototype technique is called "Wizard of Oz," after the wizard in L. Frank Baum's classic book (and the classic movie) *The Wonderful Wizard of Oz*. The wizard was actually just an ordinary person but, through the use of smoke and mirrors, he managed to appear mysterious and omnipotent. In other words, it was all a fake: the wizard had no special powers.

The Wizard of Oz method can be used to mimic a huge, powerful system long before it can be built. It can be remarkably effective in the early stages of product development. I once used this method to test a system for making airline reservations that had been designed by a research group at the Xerox Corporation's Palo Alto Research Center (today it is simply the Palo Alto Research Center, or PARC). We brought people into my laboratory in San Diego one at a time, seated them in a small, isolated room, and had them type their travel requirements into a computer. They thought they were interacting with an automated travel assistance program, but in fact, one of my graduate students was sitting in an adjacent room, reading the typed queries and typing back responses (looking up real travel schedules where appropriate). This simulation taught us a lot about the requirements for such a system. We learned, for example, that people's sentences were very different from the ones we had designed the system to handle. Example: One of the people we tested requested a round-trip ticket between San Diego and San Francisco. After the system had determined the desired flight to San Francisco, it asked, "When would you like to return?" The person responded, "I would like to leave on the following Tuesday, but I have to be back before my first class at 9 AM." We soon learned that it wasn't sufficient to understand the sentences: we also had to do problem-solving, using considerable knowledge about such things as airport and meeting locations, traffic patterns, delays for getting baggage and rental cars, and of course, parking—more than our system was capable of doing. Our initial goal was to understand language. The studies demonstrated that the goal was too limited: we needed to understand human activities.

Prototyping during the problem specification phase is done mainly to ensure that the problem is well understood. If the target population is already using something related to the new product, that can be considered a prototype. During the problem solution phase of design, then real prototypes of the proposed solution are invoked.

TESTING

Gather a small group of people who correspond as closely as possible to the target population—those for whom the product is intended. Have them use the prototypes as nearly as possible to the way they would actually use them. If the device is normally used by one person, test one person at a time. If it is normally used by a group, test a group. The only exception is that even if the normal usage is by a single person, it is useful to ask a pair of people to use it together, one person operating the prototype, the other guiding the actions and interpreting the results (aloud). Using pairs in this way causes them to discuss their ideas, hypotheses, and frustrations openly and naturally. The research team should be observing, either by sitting behind those being tested (so as not to distract them) or by watching through video in another room (but having the video camera visible and after describing the procedure). Video recordings of the tests are often quite valuable, both for later showings to team members who could not be present and for review.

When the study is over, get more detailed information about the people's thought processes by retracing their steps, reminding them of their actions, and questioning them. Sometimes it helps to show them video recordings of their activities as reminders.

How many people should be studied? Opinions vary, but my associate, Jakob Nielsen, has long championed the number five: five people studied individually. Then, study the results, refine them, and do another iteration, testing five different people. Five is usually enough to give major findings. And if you really want to test many more people, it is far more effective to do one test of five, use the results to improve the system, and then keep iterating the test-design cycle until you have tested the desired number of people. This gives multiple iterations of improvement, rather than just one.

Like prototyping, testing is done in the problem specification phase to ensure that the problem is well understood, then done again in the problem solution phase to ensure that the new design meets the needs and abilities of those who will use it.

ITERATION

The role of iteration in human-centered design is to enable continual refinement and enhancement. The goal is rapid prototyping and testing, or in the words of David Kelly, Stanford professor and cofounder of the design firm IDEO, "Fail frequently, fail fast."

Many rational executives (and government officials) never quite understand this aspect of the design process. Why would you want to fail? They seem to think that all that is necessary is to determine the requirements, then build to those requirements. Tests, they believe, are only necessary to ensure that the requirements are met. It is this philosophy that leads to so many unusable systems. Deliberate tests and modifications make things better. Failures are to be encouraged—actually, they shouldn't be called failures: they should be thought of as learning experiences. If everything works perfectly, little is learned. Learning occurs when there are difficulties.

The hardest part of design is getting the requirements right, which means ensuring that the right problem is being solved, as well as that the solution is appropriate. Requirements made in the abstract are invariably wrong. Requirements produced by asking people what they need are invariably wrong. Requirements are developed by watching people in their natural environment.

When people are asked what they need, they primarily think of the everyday problems they face, seldom noticing larger failures, larger needs. They don't question the major methods they use. Moreover, even if they carefully explain how they do their tasks and then agree that you got it right when you present it back to them, when you watch them, they will often deviate from their own description. "Why?" you ask. "Oh, I had to do this one differently," they might reply; "this was a special case." It turns out that most cases are "special." Any system that does not allow for special cases will fail.

Getting the requirements right involves repeated study and testing: iteration. Observe and study: decide what the problem might be, and use the results of tests to determine which parts of the design work, which don't. Then iterate through all four processes once again. Collect more design research if necessary, create more ideas, develop the prototypes, and test them.

With each cycle, the tests and observations can be more targeted and more efficient. With each cycle of the iteration, the ideas become clearer, the specifications better defined, and the prototypes closer approximations to the target, the actual product. After the first few iterations, it is time to start converging upon a solution. The several different prototype ideas can be collapsed into one.

When does the process end? That is up to the product manager, who needs to deliver the highest-possible quality while meeting the schedule. In product development, schedule and cost provide very strong constraints, so it is up to the design team to meet these requirements while getting to an acceptable, high-quality design. No matter how much time the design team has been allocated, the final results only seem to appear in the last twenty-four hours before the deadline. (It's like writing: no matter how much time you are given, it's finished only hours before the deadline.)

ACTIVITY-CENTERED VERSUS HUMAN-CENTERED DESIGN

The intense focus on individuals is one of the hallmarks of human-centered design, ensuring that products do fit real needs, that they are usable and understandable. But what if the product is intended for people all across the world? Many manufacturers make essentially the same product for everyone. Although automobiles are slightly modified for the requirements of a country, they are all basically the same the world round. The same is true for cameras, computers, telephones, tablets, television sets, and refrigerators. Yes, there are some regional differences, but remarkably little. Even products specifically designed for one culture—rice cookers, for example—get adopted by other cultures elsewhere.

How can we pretend to accommodate all of these very different, very disparate people? The answer is to focus on activities, not the individual person. I call this *activity-centered design.* Let the activity define the product and its structure. Let the conceptual model of the product be built around the conceptual model of the activity.

Why does this work? Because people's activities across the world tend to be similar. Moreover, although people are unwilling to learn systems that appear to have arbitrary, incomprehensible requirements, they are quite willing to learn things that appear to be essential to the activity. Does this violate the principles of human-centered design? Not at all: consider it an enhancement of HCD. After all, the activities are done by and for people. Activity-centered approaches are human-centered approaches, far better suited for large, nonhomogeneous populations.

Take another look at the automobile, basically identical all across the world. It requires numerous actions, many of which make little sense outside of the activity and that add to the complexity of driving and to the rather long period it takes to become an accomplished, skilled driver. There is the need to master foot pedals, to steer, use turn signals, control the lights, and watch the road, all while being aware of events on either side of and behind the vehicle, and perhaps while maintaining conversations with the other people in the auto. In addition, instruments on the panel need to be watched, especially the speed indicator, as well as the water temperature, oil pressure, and fuel level. The locations of the rear- and side-view mirrors require the eyes to be off the road ahead for considerable time.

People learn to drive cars quite successfully despite the need to master so many subcomponent tasks. Given the design of the car and the activity of driving, each task seems appropriate. Yes, we can make things better. Automatic transmissions eliminate the need for the third pedal, the clutch. Heads-up displays mean that critical instrument panel and navigation information can be displayed in the space in front of the driver, so no eye movements are required to monitor them (although it requires an attentional shift, which does take attention off the road). Someday we will replace the three different mirrors with one video display that shows objects on all sides of the car in one image, simplifying yet another action. How do we make things better? By careful study of the activities that go on during driving.

Support the activities while being sensitive to human capabilities, and people will accept the design and learn whatever is necessary.

ON THE DIFFERENCES BETWEEN TASKS AND ACTIVITIES

One comment: there is a difference between task and activity. I emphasize the need to design for activities: designing for tasks is usually too restrictive. An activity is a high-level structure, perhaps "go shopping." A task is a lower-level component of an activity, such as "drive to the market," "find a shopping basket," "use a shopping list to guide the purchases," and so forth.

An activity is a collected set of tasks, but all performed together toward a common high-level goal. A task is an organized, cohesive set of operations directed toward a single, low-level goal. Products have to provide support for both activities and the various tasks that are involved. Well-designed devices will package together the various tasks that are required to support an activity, making them work seamlessly with one another, making sure the work done for one does not interfere with the requirements for another.

Activities are hierarchical, so a high-level activity (going to work) will have under it numerous lower-level ones. In turn, low-level activities spawn "tasks," and tasks are eventually executed by basic "operations." The American psychologists Charles Carver and Michael Scheier suggest that goals have three fundamental levels that control activities. Be-goals are at the highest, most abstract level and govern a person's being: they determine why people act, are fundamental and long lasting, and determine one's self-image. Of far more practical concern for everyday activity is the next level down, the do-goal, which is more akin to the goal I discuss in the seven stages of activity. Do-goals determine the plans and actions to be performed for an activity. The lowest level of this hierarchy is the motor-goal, which specifies just how the actions are performed: this is more at the level of tasks and operations rather than activities. The German psychologist Marc Hassenzahl has shown how this three-level analysis can be used to guide in the development and analysis of a person's experience (the user experience, usually abbreviated UX) in interacting with products.

Focusing upon tasks is too limiting. Apple's success with its music player, the iPod, was because Apple supported the entire activity involved in listening to music: discovering it, purchasing it, getting it into the music player, developing playlists (that could be shared), and listening to the music. Apple also allowed other companies to add to the capabilities of the system with external speakers, microphones, all sorts of accessories. Apple made it possible to send the music throughout the home, to be listened to on those other companies' sound systems. Apple's success was due to its combination of two factors: brilliant design plus support for the entire activity of music enjoyment.

Design for individuals and the results may be wonderful for the particular people they were designed for, but a mismatch for others. Design for activities and the result will be usable by everyone. A major benefit is that if the design requirements are consistent with their activities, people will tolerate complexity and the requirements to learn something new: as long as the complexity and the new things to be learned feel appropriate to the task, they will feel natural and be viewed as reasonable.

ITERATIVE DESIGN VERSUS LINEAR STAGES

The traditional design process is linear, sometimes called the *waterfall method* because progress goes in a single direction, and once decisions have been made, it is difficult or impossible to go back. This is in contrast to the iterative method of human-centered design, where the process is circular, with continual refinement, continual change, and encouragement of backtracking, rethinking early decisions. Many software developers experiment with variations on the theme, variously called by such names as Scrum and Agile.

Linear, waterfall methods make logical sense. It makes sense that design research should precede design, design precede engineering development, engineering precede manufacturing, and so on. Iteration makes sense in helping to clarify the problem statement and requirements; but when projects are large, involving considerable people, time, and budget, it would be horribly expensive to allow iteration to last too long. On the other hand, proponents of iterative development have seen far too many project teams rush to develop requirements that later prove to be faulty, sometimes wasting huge amounts of money as a result. Numerous large projects have failed at a cost of multiple billions of dollars.

The most traditional waterfall methods are called *gated* methods because they have a linear set of phases or stages, with a gate blocking transition from one stage to the next. The gate is a management review during which progress is evaluated and the decision to proceed to the next stage is made.

Which method is superior? As is invariably the case where fierce debate is involved, both have virtues and both have deficits. In design, one of the most difficult activities is to get the specifications right: in other words, to determine that the correct problem is being solved. Iterative methods are designed to defer the formation of rigid specifications, to start off by diverging across a large set of possible requirements or problem statements before convergence, then again diverging across a large number of potential solutions before converging. Early prototypes have to be tested through real interaction with the target population in order to refine the requirements.

The iterative method, however, is best suited for the early design phases of a product, not for the later stages. It also has difficulty scaling its procedures to handle large projects. It is extremely difficult to deploy successfully on projects that involve hundreds or even thousands of developers, take years to complete, and cost in the millions or billions of dollars. These large projects include complex consumer goods and large programming jobs, such as automobiles; operating systems for computers, tablets, and phones; and word processors and spreadsheets.

Decision gates give management much better control over the process than they have in the iterative methods. However, they are cumbersome. The management reviews at each of the gates can take considerable time, both in preparation for them and then in the decision time after the presentations. Weeks can be wasted because of the difficulty of scheduling all the senior executives from the different divisions of the company who wish to have a say.

Many groups are experimenting with different ways of managing the product development process. The best methods combine the benefits of both iteration and stage reviews. Iteration occurs inside the stages, between the gates. The goal is to have the best of both worlds: iterative experimentation to refine the problem and the solution, coupled with management reviews at the gates.

The trick is to delay precise specification of the product requirements until some iterative testing with rapidly deployed prototypes has been done, while still keeping tight control over schedule, budget, and quality. It may appear impossible to prototype some large projects (for example, large transportation systems), but even there a lot can be done. The prototypes might be scaled objects, constructed by model makers or 3-D printing methods. Even well-rendered drawings and videos of cartoons or simple animation sketches can be useful. Virtual reality computer aids allow people to envision themselves using the final product, and in the case of a building, to envision living or working within it. All of these methods can provide rapid feedback before much time or money has been expended.

The hardest part of the development of complex products is management: organizing and communicating and synchronizing the many different people, groups, and departmental divisions that are required to make it happen. Large projects are especially difficult, not only because of the problem of managing so many different people and groups, but also because the projects' long time horizon introduces new difficulties. In the many years it takes to go from project formulation to completion, the requirements and technologies will probably change, making some of the proposed work irrelevant and obsolete; the people who will make use of the results might very well change; and the people involved in executing the project definitely will change.

Some people will leave the project, perhaps because of illness or injury, retirement or promotion. Some will change companies and others will move on to other jobs in the same company. Whatever the reason, considerable time is lost finding replacements and then bringing them up to the full knowledge and skill level required. Sometimes this is not even possible because critical knowledge about project decisions and methods are in the form we call *implicit knowledge*; that is, within the heads of the workers. When workers leave, their implicit knowledge goes with them. The management of large projects is a difficult challenge.

What I Just Told You? It Doesn't Really Work That Way

The preceding sections describe the human-centered design process for product development. But there is an old joke about the difference between theory and practice:

> *In theory, there is no difference between theory and practice.*
> *In practice, there is.*

The HCD process describes the ideal. But the reality of life within a business often forces people to behave quite differently from that ideal. One disenchanted member of the design team for a consumer products company told me that although his company professes to believe in user experience and to follow human-centered design, in practice there are only two drivers of new products:

1. Adding features to match the competition
2. Adding some feature driven by a new technology

"Do we look for human needs?" he asked, rhetorically. "No," he answered himself.

This is typical: market-driven pressures plus an engineering-driven company yield ever-increasing features, complexity, and confusion. But even companies that do intend to search for human needs are thwarted by the severe challenges of the product development process, in particular, the challenges of insufficient time and insufficient money. In fact, having watched many products succumb to these challenges, I propose a "Law of Product Development":

DON NORMAN'S LAW OF PRODUCT DEVELOPMENT

The day a product development process starts, it is behind schedule and above budget.

Product launches are always accompanied by schedules and budgets. Usually the schedule is driven by outside considerations, including holidays, special product announcement opportunities, and even factory schedules. One product I worked on was given the unrealistic timeline of four weeks because the factory in Spain would then go on vacation, and when the workers returned, it would be too late to get the product out in time for the Christmas buying season.

Moreover, product development takes time even to get started. People are never sitting around with nothing to do, waiting to be called for the product. No, they must be recruited, vetted, and then transitioned off their current jobs. This all takes time, time that is seldom scheduled.

So imagine a design team being told that it is about to work on a new product. "Wonderful," cries the team; "we'll immediately send out our design researchers to study target customers." "How long will that take?" asks the product manager. "Oh, we can do it quickly: a week or two to make the arrangements, and then two weeks in the field. Perhaps a week to distill the findings. Four or five weeks." "Sorry," says the product manager, "we don't have time. For that matter, we don't have the budget to send a team into the field for two weeks." "But it's essential if we really want to understand the customer," argues the design team. "You're absolutely right," says the product manager, "but we're behind schedule: we can't afford either the time or the money. Next time. Next time we will do it right." Except there is never a next time, because when the next time comes around, the same arguments get repeated: that product also starts behind schedule and over budget.

Product development involves an incredible mix of disciplines, from designers to engineers and programmers, manufacturing, packaging, sales, marketing, and service. And more. The product has to appeal to the current customer base as well as to expand beyond to new customers. Patents create a minefield for designers and engineers, for today it is almost impossible to design or build anything that doesn't conflict with patents, which means redesign to work one's way through the mines.

Each of the separate disciplines has a different view of the product, each has different but specific requirements to be met. Often the requirements posed by each discipline are contradictory or incompatible with those of the other disciplines. But all of them are correct when viewed from their respective perspective. In most companies, however, the disciplines work separately, design passing its results to engineering and programming, which modify the requirements to fit their needs. They then pass their results to manufacturing, which does further modification, then marketing requests changes. It's a mess.

What is the solution?

The way to handle the time crunch that eliminates the ability to do good up-front design research is to separate that process from the product team: have design researchers always out in the field, always studying potential products and customers. Then, when the product team is launched, the designers can say, "We already examined this case, so here are our recommendations." The same argument applies to market researchers.

The clash of disciplines can be resolved by multidisciplinary teams whose participants learn to understand and respect the requirements of one another. Good product development teams work as harmonious groups, with representatives from all the relevant disciplines present at all times. If all the viewpoints and requirements can be understood by all participants, it is often possible to think of creative solutions that satisfy most of the issues. Note that working with these teams is also a challenge. Everyone speaks a different technical language. Each discipline thinks it is the most important part of the process. Quite often, each discipline thinks the others are stupid, that they are making inane requests. It takes a skilled product manager to create mutual understanding and respect. But it can be done.

The design practices described by the double-diamond and the human-centered design process are the ideal. Even though the ideal can seldom be met in practice, it is always good to aim for the ideal, but to be realistic about the time and budgetary challenges. These can be overcome, but only if they are recognized and designed into the process. Multidisciplinary teams allow for enhanced communication and collaboration, often saving both time and money.

The Design Challenge

It is difficult to do good design. That is why it is such a rich, engaging profession with results that can be powerful and effective. Designers are asked to figure out how to manage complex things, to manage the interaction of technology and people. Good designers are quick learners, for today they might be asked to design a camera; tomorrow, to design a transportation system or a company's organizational structure. How can one person work across so many different domains? Because the fundamental principles of designing for people are the same across all domains. People are the same, and so the design principles are the same.

Designers are only one part of the complex chain of processes and different professions involved in producing a product. Although the theme of this book is the importance of satisfying the needs of the people who will ultimately use the product, other aspects of the product are important; for example, its engineering effectiveness, which includes its capabilities, reliability, and serviceability; its cost; and its financial viability, which usually means profitability. Will people buy it? Each of these aspects poses its own set of requirements, sometimes ones that appear to be in opposition to those of the other aspects. Schedule and budget are often the two most severe constraints.

Designers try hard to determine people's real needs and to fulfill them, whereas marketing is concerned with determining what people will actually buy. What people need and what they buy are two different things, but both are important. It doesn't matter how great the product is if nobody buys it. Similarly, if a company's products are not profitable, the company might very well go out of business. In dysfunctional companies, each division of the company is skeptical of the value added to the product by the other divisions.

In a properly run organization, team members coming from all the various aspects of the product cycle get together to share their requirements and to work harmoniously to design and produce a product that satisfies them, or at least that does so with acceptable compromises. In dysfunctional companies, each team works in isolation, often arguing with the other teams, often watching its designs or specifications get changed by others in what each team considers an unreasonable way. Producing a good product requires a lot more than good technical skills: it requires a harmonious, smoothly functioning, cooperative and respectful organization.

The design process must address numerous constraints. In the sections that follow, I examine these other factors.

PRODUCTS HAVE MULTIPLE, CONFLICTING REQUIREMENTS

Designers must please their clients, who are not always the end users. Consider major household appliances, such as stoves, refrigerators, dishwashers, and clothes washers and dryers; and even faucets and thermostats for heating and air-conditioning systems.

They are often purchased by housing developers or landlords. In businesses, purchasing departments make decisions for large companies; and owners or managers, for small companies. In all these cases, the purchaser is probably interested primarily in price, perhaps in size or appearance, almost certainly not in usability. And once devices are purchased and installed, the purchaser has no further interest in them. The manufacturer has to attend to the requirements of these decision makers, because these are the people who actually buy the product. Yes, the needs of the eventual users are important, but to the business, they seem of secondary importance.

In some situations, cost dominates. Suppose, for example, you are part of a design team for office copiers. In large companies, copying machines are purchased by the Printing and Duplicating Center, then dispersed to the various departments. The copiers are purchased after a formal "request for proposals" has gone out to manufacturers and dealers of machines. The selection is almost always based on price plus a list of required features. Usability? Not considered. Training costs? Not considered. Maintenance? Not considered. There are no requirements regarding understandability or usability of the product, even though in the end those aspects of the product can end up costing the company a lot of money in wasted time, increased need for service calls and training, and even lowered staff morale and lower productivity.

The focus on sales price is one reason we get unusable copying machines and telephone systems in our places of employment. If people complained strongly enough, usability could become a requirement in the purchasing specifications, and that requirement could trickle back to the designers. But without this feedback, designers must often design the cheapest possible products because those are what sell. Designers need to understand their customers, and in many cases, the customer is the person who purchases the product, not the person who actually uses it. It is just as important to study those who do the purchasing as it is to study those who use it.

To make matters even more difficult, yet another set of people needs to be considered: the engineers, developers, manufacturing, services, sales, and marketing people who have to translate the ideas from the design team into reality, and then sell and support the product after it is shipped. These groups are users, too, not of the product itself, but of the output of the design team. Designers are used to accommodating the needs of the product users, but they seldom consider the needs of the other groups involved in the product process. But if their needs are not considered, then as the product development moves through the process from design to engineering, to marketing, to manufacturing, and so on, each new group will discover that it doesn't meet their needs, so they will change it. But piecemeal, after-the-fact changes invariably weaken the cohesion of the product. If all these requirements were known at the start of the design process, a much more satisfactory resolution could have been devised.

Usually the different company divisions have intelligent people trying to do what is best for the company. When they make changes to a design, it is because their requirements were not suitably served. Their concerns and needs are legitimate, but changes introduced in this way are almost always detrimental. The best way to counteract this is to ensure that representatives from all the divisions are present during the entire design process, starting with the decision to launch the product, continuing all the way through shipment to customers, service requirements, and repairs and returns. This way, all the concerns can be heard as soon as they are discovered. There must be a multidisciplinary team overseeing the entire design, engineering, and manufacturing process that shares all departmental issues and concerns from day one, so that everyone can design to satisfy them, and when conflicts arise, the group together can determine the most satisfactory solution. Sadly, it is the rare company that is organized this way.

Design is a complex activity. But the only way this complex process comes together is if all the relevant parties work together as a team. It isn't design against engineering, against marketing, against manufacturing: it is design together with all these other players. Design must take into account sales and marketing, servicing and help desks, engineering and manufacturing, costs and schedules. That's why it's so challenging. That's why it's so much fun and rewarding when it all comes together to create a successful product.

DESIGNING FOR SPECIAL PEOPLE

There is no such thing as the average person. This poses a particular problem for the designer, who usually must come up with a single design for everyone. The designer can consult handbooks with tables that show average arm reach and seated height, how far the average person can stretch backward while seated, and how much room is needed for average hips, knees, and elbows. *Physical anthropometry* is what the field is called. With data, the designer can try to meet the size requirements for almost everyone, say for the 90th, 95th, or even the 99th percentile. Suppose the product is designed to accommodate the 95th percentile, that is, for everyone except the 5 percent of people who are smaller or larger. That leaves out a lot of people. The United States has approximately 300 million people, so 5 percent is 15 million. Even if the design aims at the 99th percentile it would still leave out 3 million people. And this is just for the United States: the world has 7 billion people. Design for the 99th percentile of the world and 70 million people are left out.

Some problems are not solved by adjustments or averages: Average a left-hander with a right-hander and what do you get? Sometimes it is simply impossible to build one product that accommodates everyone, so the answer is to build different versions of the product. After all, we would not be happy with a store that sells only one size and type of clothing: we expect clothing that fits our bodies, and people come in a very wide range of sizes. We don't expect the large variety of goods found in a clothing store to apply to all people or activities; we expect a wide variety of cooking appliances, automobiles, and tools so we can select the ones that precisely match our requirements. One device simply cannot work for everyone. Even such simple tools as pencils need to be designed differently for different activities and types of people.

Consider the special problems of the aged and infirm, the handicapped, the blind or near blind, the deaf or hard of hearing, the very short or very tall, or people who speak other languages. Design for interests and skill levels. Don't be trapped by overly general, inaccurate stereotypes. I return to these groups in the next section.

THE STIGMA PROBLEM

> *"I don't want to go into a care facility. I'd have to be around all those old people."* (Comment by a 95-year-old man.)

Many devices designed to aid people with particular difficulties fail. They may be well designed, they may solve the problem, but they are rejected by their intended users. Why? Most people do not wish to advertise their infirmities. Actually, many people do not wish to admit having infirmities, even to themselves.

When Sam Farber wanted to develop a set of household tools that his arthritic wife could use, he worked hard to find a solution that was good for everyone. The result was a series of tools that revolutionized this field. For example, vegetable peelers used to be an inexpensive, simple metal tool, often of the form shown on the left in Figure 6.3. These were awkward to use, painful to hold, and not even that effective at peeling, but everyone assumed that this was how they had to be.

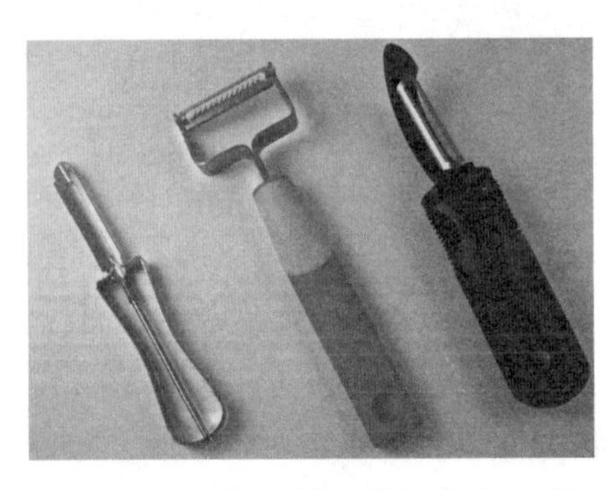

FIGURE 6.3. Three Vegetable Peelers. The traditional metal vegetable peeler is shown on the left: inexpensive, but uncomfortable. The OXO peeler that revolutionized the industry is shown on the right. The result of this revolution is shown in the middle, a peeler from the Swiss company Kuhn Rikon: colorful and comfortable.

After considerable research, Farber settled upon the peeler shown on the right in Figure 6.3 and built a company, OXO, to manufacture and distribute it. Even though the peeler was designed for someone with arthritis, it was advertised as a better peeler for everyone. It was. Even though the design was more expensive than the regular peeler, it was so successful that today, many companies make variations on this theme. You may have trouble seeing the OXO peeler as revolutionary because today, many have followed in these footsteps. Design has become a major theme for even simple tools such as peelers, as demonstrated by the center peeler of Figure 6.3.

Consider the two things special about the OXO peeler: cost and design for someone with an infirmity. Cost? The original peeler was very inexpensive, so a peeler that is many times the cost of the inexpensive one is still inexpensive. What about the special design for people with arthritis? The virtues for them were never mentioned, so how did they find it? OXO did the right thing and let the world know that this was a better product. And the world took note and made it successful. As for people who needed the better handle? It didn't take long for the word to spread. Today, many companies have followed the OXO route, producing peelers that work extremely well, are comfortable, and are colorful. See Figure 6.3.

Would you use a walker, wheelchair, crutches, or a cane? Many people avoid these, even though they need them, because of the negative image they cast: the stigma. Why? Years ago, a cane was fashionable: people who didn't need them would use them anyway, twirling them, pointing with them, hiding brandy or whisky, knives or guns inside their handles. Just look at any movie depicting nineteenth-century London. Why can't devices for those who need them be as sophisticated and fashionable today?

Of all the devices intended to aid the elderly, perhaps the most shunned is the walker. Most of these devices are ugly. They cry out, "Disability here." Why not transform them into products to be proud of? Fashion statements, perhaps. This thinking has already begun with some medical appliances. Some companies are making hearing aids and glasses for children and adolescents with special colors and styles that appeal to these age groups. Fashion accessories. Why not?

Those of you who are young, do not smirk. Physical disabilities may begin early, starting in the midtwenties. By their midforties, most people's eyes can no longer adjust sufficiently to focus over the entire range of distances, so something is necessary to compensate, whether reading glasses, bifocals, special contact lenses, or even surgical correction.

Many people in their eighties and nineties are still in good mental and physical shape, and the accumulated wisdom of their years leads to superior performance in many tasks. But physical strength and agility do decrease, reaction time slows, and vision and hearing show impairments, along with decreased ability to divide attention or switch rapidly among competing tasks.

For anyone who is considering growing old, I remind you that although physical abilities diminish with age, many mental capacities continue to improve, especially those dependent upon an expert accumulation of experience, deep reflection, and enhanced knowledge. Younger people are more agile, more willing to experiment and take risks. Older people have more knowledge and wisdom. The world benefits from having a mix and so do design teams.

Designing for people with special needs is often called *inclusive* or *universal design*. Those names are fitting, for it is often the case that everyone benefits. Make the lettering larger, with high-contrast type, and everyone can read it better. In dim light, even the people with the world's best eyesight will benefit from such lettering. Make things adjustable, and you will find that more people can use it, and even people who liked it before may now like it better. Just as I invoke the so-called error message of Figure 4.6 as my normal way of exiting a program because it is easier than the so-called correct way, special features made for people with special needs often turn out to be useful for a wide variety of people.

The best solution to the problem of designing for everyone is flexibility: flexibility in the size of the images on computer screens, in the sizes, heights, and angles of tables and chairs. Allow people to adjust their own seats, tables, and working devices. Allow them to adjust lighting, font size, and contrast. Flexibility on our highways might mean ensuring that there are alternative routes with different speed limits. Fixed solutions will invariably fail with some people; flexible solutions at least offer a chance for those with different needs.

Complexity Is Good; It Is Confusion That Is Bad

The everyday kitchen is complex. We have multiple instruments just for serving and eating food. The typical kitchen contains all sorts of cutting utensils, heating units, and cooking apparatus. The easiest way to understand the complexity is to try to cook in an unfamiliar kitchen. Even excellent cooks have trouble working in a new environment.

Someone else's kitchen looks complicated and confusing, but your own kitchen does not. The same can probably be said for every room in the home. Notice that this feeling of confusion is really one of knowledge. My kitchen looks confusing to you, but not to me. In turn, your kitchen looks confusing to me, but not to you. So the confusion is not in the kitchen: it is in the mind. "Why can't things be made simple?" goes the cry. Well, one reason is that life is complex, as are the tasks we encounter. Our tools must match the tasks.

I feel so strongly about this that I wrote an entire book on the topic, *Living with Complexity,* in which I argued that complexity is essential: it is confusion that is undesirable. I distinguished between "complexity," which we need to match the activities we take part in, and "complicated," which I defined to mean "confusing." How do we avoid confusion? Ah, here is where the designer's skills come into play.

The most important principle for taming complexity is to provide a good conceptual model, which has already been well covered in this book. Remember the kitchen's apparent complexity? The people who use it understand why each item is stored where it is: there is usually structure to the apparent randomness. Even exceptions fit: even if the reason is something like, "It was too big to fit in the proper drawer and I didn't know where else to put it," that is reason enough to give structure and understanding to the person who stored the item there. Complex things are no longer complicated once they are understood.

Standardization and Technology

If we examine the history of advances in all technological fields, we see that some improvements come naturally through the technology itself, others come through standardization. The early history of the automobile is a good example. The first cars were very difficult to operate. They required strength and skill beyond the abilities of many. Some problems were solved through automation: the choke, the spark advance, and the starter engine. Other aspects of cars and driving were standardized through the long process of international standards committees:

- On which side of the road to drive (constant within a country, but variable across countries)
- On which side of the car the driver sits (depends upon which side of the road the car is driven)
- The location of essential components: steering wheel, brake, clutch, and accelerator (the same, whether on the left- or right-hand side of the car)

Standardization is one type of cultural constraint. With standardization, once you have learned to drive one car, you feel justifiably confident that you can drive any car, anyplace in the world. Standardization provides a major breakthrough in usability.

ESTABLISHING STANDARDS

I have enough friends on national and international standards committees to realize that the process of determining an internationally accepted standard is laborious. Even when all parties agree on the merits of standardization, the task of selecting standards becomes a lengthy, politicized issue. A small company can standardize its products without too much difficulty, but it is much more difficult for an industrial, national, or international body to agree to standards. There even exists a standardized procedure for establishing national and international standards. A set of national and international organizations works on standards; when a new standard is proposed, it must work its way through the organizational hierarchy. Each step is complex, for if there are three ways of doing something, then there are sure to be strong proponents of each of the three ways, plus people who will argue that it is too early to standardize.

FIGURE 6.4. The Nonstandard Clock. What time is it? This clock is just as logical as the standard one, except the hands move in the opposite direction and "12" is not in its usual place. Same logic, though. So why is it so difficult to read? What time is being displayed? 7:11, of course.

Each proposal is debated at the standards committee meeting where it is presented, then taken back to the sponsoring organization—which is sometimes a company, sometimes a professional society—where objections and counter-objections are collected. Then the standards committee meets again to discuss the objections. And again and again and again. Any company that is already marketing a product that meets the proposed standard will have a huge economic advantage, and the debates are therefore often affected as much by the economics and politics of the issues as by real technological substance. The process is almost guaranteed to take five years, and quite often longer.

The resulting standard is usually a compromise among the various competing positions, oftentimes an inferior compromise. Sometimes the answer is to agree on several incompatible standards. Witness the existence of both metric and English units; of left-hand- and right-hand-drive automobiles. There are several international standards for the voltages and frequencies of electricity, and several different kinds of electrical plugs and sockets—which cannot be interchanged.

WHY STANDARDS ARE NECESSARY: A SIMPLE ILLUSTRATION

With all these difficulties and with the continual advances in technology, are standards really necessary? Yes, they are. Take the everyday clock. It's standardized. Consider how much trouble you would have telling time with a backward clock, where the hands revolved "counterclockwise." A few such clocks exist, primarily as humorous conversation pieces. When a clock truly violates standards, such as the one in Figure 6.4 on the previous page, it is difficult to determine what time is being displayed. Why? The logic behind the time display is identical to that of conventional clocks: there are only two differences—the hands rotate in the opposite direction (counterclockwise) and the location of "12," usually at the top, has been moved. This clock is just as logical as the standard one. It bothers us because we have standardized on a different scheme, on the very definition of the term *clockwise.* Without such standardization, clock reading would be more difficult: you'd always have to figure out the mapping.

A STANDARD THAT TOOK SO LONG, TECHNOLOGY OVERRAN IT

I myself participated at the very end of the incredibly long, complex political process of establishing the US standards for high-definition television. In the 1970s, the Japanese developed a national television system that had much higher resolution than the standards then in use: they called it "high-definition television."

In 1995, two decades later, the television industry in the United States proposed its own high-definition TV standard (HDTV) to the Federal Communications Commission (FCC). But the computer industry pointed out that the proposals were not compatible with the way that computers displayed images, so the FCC objected to the proposed standards. Apple mobilized other members of the industry and, as vice president of advanced technology, I was selected to be the spokesperson for Apple. (In the following description, ignore the jargon—it doesn't matter.) The TV industry proposed a wide variety of permissible formats, including ones with rectangular pixels and interlaced scan. Because of the technical limitations in the 1990s, it was suggested that the highest-quality picture have 1,080 interlaced lines (1080i). We wanted only progressive scan, so we insisted upon 720 lines, progressively displayed (720p), arguing that the progressive nature of the scan made up for the lesser number of lines.

The battle was heated. The FCC told all the competing parties to lock themselves into a room and not to come out until they had reached agreement. As a result, I spent many hours in lawyers' offices. We ended up with a crazy agreement that recognized multiple variations of the standard, with resolutions of 480i and 480p (called *standard definition*), 720p and 1080i (called *high-definition*), and two different aspect ratios for the screens (the ratio of width to height), 4:3 (= 1.3)—the old standard—and 16:9 (= 1.8)—the new standard. In addition, a large number of frame rates were supported (basically, how many times per second the image was transmitted). Yes, it was a standard, or more accurately a large number of standards. In fact, one of the allowed methods of transmission was to use any method (as long as it carried its own specifications along with the signal). It was a mess, but we did reach agreement. After the standard was made official in 1996, it took roughly ten more years for HDTV to become accepted, helped, finally, by a new generation of television displays that were large, thin, and inexpensive. The whole process took roughly thirty-five years from the first broadcasts by the Japanese.

Was it worth the fight? Yes and no. In the thirty-five years that it took to reach the standard, the technology continued to evolve, so the resulting standard was far superior to the first one proposed so many years before. Moreover, the HDTV of today is a huge improvement over what we had before (now called "standard definition"). But the minutiae of details that were the focus of the fight between the computer and TV companies was silly. My technical experts continually tried to demonstrate to me the superiority of 720p images over 1080i, but it took me hours of viewing special scenes under expert guidance to see the deficiencies of the interlaced images (the differences only show up with complex moving images). So why did we care?

Television displays and compression techniques have improved so much that interlacing is no longer needed. Images at 1080p, once thought to be impossible, are now commonplace. Sophisticated algorithms and high-speed processors make it possible to transform one standard into another; even rectangular pixels are no longer a problem.

As I write these words, the main problem is the discrepancy in aspect ratios. Movies come in many different aspect ratios (none of them the new standard) so when TV screens show movies, they either have to cut off part of the image or leave parts of the screen black. Why was the HDTV aspect ratio set at 16:9 (or 1.8) if no movies used that ratio? Because engineers liked it: square the old aspect ratio of 4:3 and you get the new one, 16:9.

Today we are about to embark on yet another standards fight over TV. First, there is three-dimensional TV: 3-D. Then there are proposals for ultra-high definition: 2,160 lines (and a doubling of the horizontal resolution as well): four times the resolution of our best TV today (1080p). One company wants eight times the resolution, and one is proposing an aspect ratio of 21:9 (= 2.3). I have seen these images and they are marvelous, although they only matter with large screens (at least 60 inches, or 1.5 meters, in diagonal length), and when the viewer is close to the display.

Standards can take so long to be established that by the time they do come into wide practice, they can be irrelevant. Nonetheless, standards are necessary. They simplify our lives and make it possible for different brands of equipment to work together in harmony.

A STANDARD THAT NEVER CAUGHT ON: DIGITAL TIME

Standardize and you simplify lives: everyone learns the system only once. But don't standardize too soon; you may be locked into a primitive technology, or you may have introduced rules that turn out to be grossly inefficient, even error-inducing. Standardize too late, and there may already be so many ways of doing things that no international standard can be agreed on. If there is agreement on an old-fashioned technology, it may be too expensive for everyone to change to the new standard. The metric system is a good example: it is a far simpler and more usable scheme for representing distance, weight, volume, and temperature than the older English system of feet, pounds, seconds, and degrees on the Fahrenheit scale. But industrial nations with a heavy commitment to the old measurement standard claim they cannot afford the massive costs and confusion of conversion. So we are stuck with two standards, at least for a few more decades.

Would you consider changing how we specify time? The current system is arbitrary. The day is divided into twenty-four rather arbitrary but standard units—hours. But we tell time in units of twelve, not twenty-four, so there have to be two cycles of twelve hours each, plus the special convention of a.m. and p.m. so we know which cycle we are talking about. Then we divide each hour into sixty minutes and each minute into sixty seconds.

What if we switched to metric divisions: seconds divided into tenths, milliseconds, and microseconds? We would have days, millidays, and microdays. There would have to be a new hour, minute, and second: call them the digital hour, the digital minute, and the digital second. It would be easy: ten digital hours to the day, one hundred digital minutes to the digital hour, one hundred digital seconds to the digital minute.

Each digital hour would last exactly 2.4 times an old hour: 144 old minutes. So the old one-hour period of the schoolroom or television program would be replaced with a half-digital hour period, or 50 digital minutes—only 20 percent longer than the current hour. We could adapt to the differences in durations with relative ease.

What do I think of it? I much prefer it. After all, the decimal system, the basis of most of the world's use of numbers and arithmetic, uses base 10 arithmetic and, as a result, arithmetic operations are much simpler in the metric system. Many societies have used other systems, 12 and 60 being common. Hence twelve for the number of items in a dozen, inches in a foot, hours in a day, and months in a year; sixty for the number of seconds in a minute, seconds in a degree, and minutes in an hour.

The French proposed that time be made into a decimal system in 1792, during the French Revolution, when the major shift to the metric system took place. The metric system for weights and lengths took hold, but not for time. Decimal time was used long enough for decimal clocks to be manufactured, but it eventually was discarded. Too bad. It is very difficult to change well-established habits. We still use the QWERTY keyboard, and the United States still measures things in inches and feet, yards and miles, Fahrenheit, ounces, and pounds. The world still measures time in units of 12 and 60, and divides the circle into 360 degrees.

In 1998, Swatch, the Swiss watch company, made its own attempt to introduce decimal time through what it called "Swatch International Time." Swatch divided the day into 1,000 ".beats," each .beat being slightly less than 90 seconds (each .beat corresponds to one digital minute). This system did not use time zones, so people the world over would be in synchrony with their watches. This does not simplify the problem of synchronizing scheduled conversations, however, because it would be difficult to get the sun to behave properly. People would still wish to wake up around sunrise, and this would occur at different Swatch times around the world. As a result, even though people would have their watches synchronized, it would still be necessary to know when they woke up, ate, went to and from work, and went to sleep, and these times would vary around the world. It isn't clear whether Swatch was serious with its proposal or whether it was one huge advertising stunt. After a few years of publicity, during which the company manufactured digital watches that told the time in .beats, it all fizzled away.

Speaking of standardization, Swatch called its basic time unit a ".beat" with the first character being a period. This nonstandard spelling wreaks havoc on spelling correction systems that aren't set up to handle words that begin with punctuation marks.

Deliberately Making Things Difficult

How can good design (design that is usable and understandable) be balanced with the need for "secrecy" or privacy, or protection? That is, some applications of design involve areas that are sensitive and necessitate strict control over who uses and understands them. Perhaps we don't want any user-in-the-street to understand enough of a system to compromise its security. Couldn't it be argued that some things shouldn't be designed well? Can't things be left cryptic, so that only those who have clearance, extended education, or whatever, can make use of the system? Sure, we have passwords, keys, and other types of security checks, but this can become wearisome for the privileged user. It appears that if good design is not ignored in some contexts, the purpose for the existence of the system will be nullified. (A computer mail question sent to me by a student, Dina Kurktchi. It is just the right question.)

In Stapleford, England, I came across a school door that was very difficult to open, requiring simultaneous operation of two latches, one at the very top of the door, the other down low. The latches were difficult to find, to reach, and to use. But the difficulties were deliberate. This was good design. The door was at a school for handicapped children, and the school didn't want the children to be able to get out to the street without an adult. Only adults were large enough to operate the two latches. Violating the rules of ease of use is just what was needed.

Most things are intended to be easy to use, but aren't. But some things are deliberately difficult to use—and ought to be. The number of things that should be difficult to use is surprisingly large:

- Any door designed to keep people in or out.
- Security systems, designed so that only authorized people will be able to use them.
- Dangerous equipment, which should be restricted.
- Dangerous operations that might lead to death or injury if done accidentally or in error.

- Secret doors, cabinets, and safes: you don't want the average person even to know that they are there, let alone to be able to work them.
- Cases deliberately intended to disrupt the normal routine action (as discussed in Chapter 5). Examples include the acknowledgment required before permanently deleting a file from a computer, safeties on pistols and rifles, and pins in fire extinguishers.
- Controls that require two simultaneous actions before the system will operate, with the controls separated so that it takes two people to work them, preventing a single person from doing an unauthorized action (used in security systems or safety-critical operations).
- Cabinets and bottles for medications and dangerous substances deliberately made difficult to open to keep them secure from children.
- Games, a category in which designers deliberately flout the laws of understandability and usability. Games are meant to be difficult; in some games, part of the challenge is to figure out what is to be done, and how.

Even where a lack of usability or understandability is deliberate, it is still important to know the rules of understandable and usable design, for two reasons. First, even deliberately difficult designs aren't entirely difficult. Usually there is one difficult part, designed to keep unauthorized people from using the device; the rest of it should follow the normal principles of good design. Second, even if your job is to make something difficult to do, you need to know how to go about doing it. In this case, the rules are useful, for they state in reverse just how to go about the task. You could systematically violate the rules like this:

- Hide critical components: make things invisible.
- Use unnatural mappings for the execution side of the action cycle, so that the relationship of the controls to the things being controlled is inappropriate or haphazard.
- Make the actions physically difficult to do.
- Require precise timing and physical manipulation.
- Do not give any feedback.

- Use unnatural mappings for the evaluation side of the action cycle, so that system state is difficult to interpret.

Safety systems pose a special problem in design. Oftentimes, the design feature added to ensure safety eliminates one danger, only to create a secondary one. When workers dig a hole in a street, they must put up barriers to prevent cars and people from falling into the hole. The barriers solve one problem, but they themselves pose another danger, often mitigated by adding signs and flashing lights to warn of the barriers. Emergency doors, lights, and alarms must often be accompanied by warning signs or barriers that control when and how they can be used.

Design: Developing Technology for People

Design is a marvelous discipline, bringing together technology and people, business and politics, culture and commerce. The different pressures on design are severe, presenting huge challenges to the designer. At the same time, the designers must always keep foremost in mind that the products are to be used by people. This is what makes design such a rewarding discipline: On the one hand, woefully complex constraints to overcome; on the other hand, the opportunity to develop things that assist and enrich the lives of people, that bring benefits and enjoyment.

CHAPTER SEVEN

DESIGN IN THE WORLD OF BUSINESS

The realities of the world impose severe constraints upon the design of products. Up to now I have described the ideal case, assuming that human-centered design principles could be followed in a vacuum; that is, without attention to the real world of competition, costs, and schedules. Conflicting requirements will come from different sources, all of which are legitimate, all of which need to be resolved. Compromises must be made by all involved.

Now it is time to examine the concerns outside of human-centered design that affect the development of products. I start with the impact of competitive forces that drive the introduction of extra features, often to excess: the cause of the disease dubbed "featuritis," whose major symptom is "creeping featurism." From there, I examine the drivers of change, starting with technological drivers. When new technologies emerge, there is a temptation to develop new products immediately. But the time for radically new products to become successful is measured in years, decades, or in some instances centuries. This causes me to examine the two forms of product innovation relevant to design: incremental (less glamorous, but most common) and radical (most glamorous, but rarely successful).

I conclude with reflections about the history and future prospects of this book. The first edition of this book has had a long and fruitful life. Twenty-five years is an amazingly long time for a book centered around technology to have remained relevant. If this revised and expanded edition lasts an equally long time, that means fifty years of *The Design of Everyday Things*. In these next twenty-five years, what new developments will take place? What will be the role of technology in our lives, for the future of books, and what are the moral obligations of the design profession? And finally, for how long will the principles in this book remain relevant? It should be no surprise that I believe they will always be just as relevant as they were twenty-five years ago, just as relevant as they are today. Why? The reason is simple. The design of technology to fit human needs and capabilities is determined by the psychology of people. Yes, technologies may change, but people stay the same.

Competitive Forces

Today, manufacturers around the world compete with one another. The competitive pressures are severe. After all, there are only a few basic ways by which a manufacturer can compete: three of the most important being price, features, and quality—unfortunately often in that order of importance. Speed is important, lest some other company get ahead in the rush for market presence. These pressures make it difficult to follow the full, iterative process of continual product improvement. Even relatively stable home products, such as automobiles, kitchen appliances, television sets, and computers, face the multiple forces of a competitive market that encourage the introduction of changes without sufficient testing and refinement.

Here is a simple, real example. I am working with a new startup company, developing an innovative line of cooking equipment. The founders had some unique ideas, pushing the technology of cooking far ahead of anything available for homes. We did numerous field tests, built numerous prototypes, and engaged a world-class industrial designer. We modified the original product concept several times, based on early feedback from potential users and advice from industry experts. But just as we were about to commission the first production of a few hand-tooled working prototypes that could be shown to potential investors and customers (an expensive proposition for the small self-funded company), other companies started displaying similar concepts in the trade shows. What? Did they steal the ideas? No, it's what is called the *Zeitgeist*, a German word meaning "spirit of the time." In other words, the time was ripe, the ideas were "in the air." The competition emerged even before we had delivered our first product. What is a small, startup company to do? It doesn't have money to compete with the large companies. It has to modify its ideas to keep ahead of the competition and come up with a demonstration that excites potential customers and wows potential investors and, more importantly, potential distributors of the product. It is the distributors who are the real customers, not the people who eventually buy the product in stores and use it in their homes. The example illustrates the real business pressures on companies: the need for speed, the concern about costs, the competition that may force the company to change its offerings, and the need to satisfy several classes of customers—investors, distributors, and, of course, the people who will actually use the product. Where should the company focus its limited resources? More user studies? Faster development? New, unique features?

The same pressures that the startup faced also impact established companies. But they have other pressures as well. Most products have a development cycle of one to two years. In order to bring out a new model every year, the design process for the new model has to have started even before the previous model has been released to customers. Moreover, mechanisms for collecting and feeding back the experiences of customers seldom exist. In an earlier era, there was close coupling between designers and users. Today, they are separated by barriers. Some companies prohibit designers from working with customers, a bizarre and senseless restriction. Why would they do this? In part to prevent leaks of the new developments to the competition, but also in part because customers may stop purchasing the current offerings if they are led to believe that a new, more advanced item is soon to come. But even where there are no such restrictions, the complexity of large organizations coupled with the relentless pressure to finish the product makes this interaction difficult. Remember Norman's Law of Chapter 6: The day a product development process starts, it is behind schedule and above budget.

FEATURITIS: A DEADLY TEMPTATION

In every successful product there lurks the carrier of an insidious disease called "featuritis," with its main symptom being "creeping featurism." The disease seems to have been first identified and named in 1976, but its origins probably go back to the earliest technologies, buried far back in the eons prior to the dawn of history. It seems unavoidable, with no known prevention. Let me explain.

Suppose we follow all the principles in this book for a wonderful, human-centered product. It obeys all design principles. It overcomes people's problems and fulfills some important needs. It is attractive and easy to use and understand. As a result, suppose the product is successful: many people buy it and tell their friends to buy it. What could be wrong with this?

The problem is that after the product has been available for a while, a number of factors inevitably appear, pushing the company toward the addition of new features—toward creeping featurism. These factors include:

- Existing customers like the product, but express a wish for more features, more functions, more capability.
- A competing company adds new features to its products, producing competitive pressures to match that offering, but to do even more in order to get ahead of the competition.
- Customers are satisfied, but sales are declining because the market is saturated: everyone who wants the product already has it. Time to add wonderful enhancements that will cause people to want the new model, to upgrade.

Featuritis is highly infectious. New products are invariably more complex, more powerful, and different in size than the first release of a product. You can see that tension playing out in music players, mobile phones, and computers, especially on smart phones, tablets, and pads. Portable devices get smaller and smaller with each release, despite the addition of more and more features (making them ever more difficult to operate). Some products, such as automobiles, home refrigerators, television sets, and kitchen stoves, also increase in complexity with each release, getting larger and more powerful.

But whether the products get larger or smaller, each new edition invariably has more features than the previous one. Featuritis is an insidious disease, difficult to eradicate, impossible to vaccinate against. It is easy for marketing pressures to insist upon the addition of new features, but there is no call—or for that matter, budget—to get rid of old, unneeded ones.

How do you know when you have encountered featuritis? By its major symptom: creeping featurism. Want an example? Look at Figure 7.1, which illustrates the changes that have overcome the simple Lego motorcycle since my first encounter with it for the first edition of this book. The original motorcycle (Figure 4.1 and Figure 7.1A) had only fifteen components and could be put together without any instructions: it had sufficient constraints that every piece had a unique location and orientation. But now, as Figure 7.1B shows, the same motorcycle has become bloated, with twenty-nine pieces. I needed instructions.

Creeping featurism is the tendency to add to the number of features of a product, often extending the number beyond all reason. There is no way that a product can remain usable and understandable by the time it has all of those special-purpose features that have been added in over time.

In her book *Different*, Harvard professor Youngme Moon argues that it is this attempt to match the competition that causes all products to be the same. When companies try to increase sales by matching every feature of their competitors, they end up hurting themselves. After all, when products from two companies match

A.

B.

FIGURE 7.1. Featuritis Strikes Lego. Figure A shows the original Lego Motorcycle available in 1988 when I used it in the first edition of this book (on the left), next to the 2013 version (on the right). The old version had only fifteen pieces. No manual was needed to put it together. For the new version, the box proudly proclaims "29 pieces." I could put the original version together without instructions. Figure B shows how far I got with the new version before I gave up and had to consult the instruction sheet. Why did Lego believe it had to change the motorcycle? Perhaps because featuritis struck real police motorcycles, causing them to increase in size and complexity and Lego felt that its toy needed to match the world. (Photographs by the author.)

feature by feature, there is no longer any reason for a customer to prefer one over another. This is competition-driven design. Unfortunately, the mind-set of matching the competitor's list of features pervades many organizations. Even if the first versions of a product are well done, human-centered, and focused upon real needs, it is the rare organization that is content to let a good product stay untouched.

Most companies compare features with their competition to determine where they are weak, so they can strengthen those areas. Wrong, argues Moon. A better strategy is to concentrate on areas where they are stronger and to strengthen them even more. Then focus all marketing and advertisements to point out the strong points. This causes the product to stand out from the mindless herd. As for the weaknesses, ignore the irrelevant ones, says Moon. The lesson is simple: don't follow blindly; focus on strengths, not weaknesses. If the product has real strengths, it can afford to just be "good enough" in the other areas.

Good design requires stepping back from competitive pressures and ensuring that the entire product be consistent, coherent, and understandable. This stance requires the leadership of the company to withstand the marketing forces that keep begging to add this feature or that, each thought to be essential for some market segment. The best products come from ignoring these competing voices and instead focusing on the true needs of the people who use the product.

Jeff Bezos, the founder and CEO of Amazon.com, calls his approach "customer obsessed." Everything is focused upon the requirements of Amazon's customers. The competition is ignored, the traditional marketing requirements are ignored. The focus is on simple, customer-driven questions: what do the customers want; how can their needs best be satisfied; what can be done better to enhance customer service and customer value? Focus on the customer, Bezos argues, and the rest takes care of itself. Many companies claim to aspire to this philosophy, but few are able to follow it. Usually it is only possible where the head of the company, the CEO, is also the founder. Once the company passes control to others, especially those who follow the traditional MBA dictum of putting profit above customer concerns, the story goes downhill. Profits may indeed increase in the short term, but eventually the product quality deteriorates to the point where customers desert. Quality only comes about by continual focus on, and attention to, the people who matter: customers.

New Technologies Force Change

Today, we have new requirements. We now need to type on small, portable devices that don't have room for a full keyboard. Touch- and gesture-sensitive screens allow a new form of typing. We can bypass typing altogether through handwriting recognition and speech understanding.

Consider the four products shown in Figure 7.2. Their appearance and methods of operations changed radically in their century of existence. Early telephones, such as the one in Figure 7.2A, did not have keyboards: a human operator intervened to make the connections. Even when operators were first replaced by automatic switching systems, the "keyboard" was a rotary dial with ten holes, one for each digit. When the dial was replaced with pushbutton keys, it suffered a slight case of featuritis: the ten positions of the dial were replaced with twelve keys: the ten digits plus * and #.

But much more interesting is the merger of devices. The human computer gave rise to laptops, small portable computers. The telephone moved to small, portable cellular phones (called mobiles in much of the world). Smart phones had large, touch-sensitive screens, operated by gesture. Soon computers merged into tablets, as did cell phones. Cameras merged with cell phones. Today, talking, video conferences, writing, photography (both still and video), and collaborative interaction of all sorts are increasingly

FIGURE 7.2. **100 Years of Telephones and Keyboards.** Figures A and B show the change in the telephone from the Western Electric crank telephone of the 1910s, where rotating the crank on the right generated a signal alerting the operator, to the phone of the 2010s. They seem to have nothing in common. Figures C and D contrast a keyboard of the 1910s with one from the 2010s. The keyboards are still laid out in the same way, but the first requires physical depression of each key; the second, a quick tracing of a finger over the relevant letters (the image shows the word *many* being entered). Credits: A, B, and C: photographs by the author; objects in A and C courtesy of the Museum of American Heritage, Palo Alto, California. D shows the "Swype" keyboard from Nuance. Image being used courtesy of Nuance Communications, Inc.

being done by one single device, available with a large variety of screen sizes, computational power, and portability. It doesn't make sense to call them computers, phones, or cameras: we need a new name. Let's call them "smart screens." In the twenty-second century, will we still have phones? I predict that although we will still talk with one another over a distance, we will not have any device called a telephone.

As the pressures for larger screens forced the demise of physical keyboards (despite the attempt to make tiny keyboards, operated with single fingers or thumbs), the keyboards were displayed on the screen whenever needed, each letter tapped one at a time. This is slow, even when the system tries to predict the word being typed so that keying can stop as soon as the correct word shows up. Several systems were soon developed that allowed the finger or stylus to trace a path among the letters of the word: word-gesture systems. The gestures were sufficiently different from one another that it wasn't even necessary to touch all the letters—it only mattered that the pattern generated by the approximation to the correct path was close enough to the desired one. This turns out to be a fast and easy way to type (Figure 7.2D).

With gesture-based systems, a major rethinking is possible. Why keep the letters in the same QWERTY arrangement? The pattern generation would be even faster if letters were rearranged to maximize speed when using a single finger or stylus to trace out the letters. Good idea, but when one of the pioneers in developing this technique, Shumin Zhai, then at IBM, tried it, he ran into the legacy problem. People knew QWERTY and balked at having to learn a different organization. Today, the word-gesture method of typing is widely used, but with QWERTY keyboards (as in Figure 7.2D).

Technology changes the way we do things, but fundamental needs remain unchanged. The need for getting thoughts written down, for telling stories, doing critical reviews, or writing fiction and nonfiction will remain. Some will be written using traditional keyboards, even on new technological devices, because the keyboard still remains the fastest way to enter words into a system, whether it be paper or electronic, physical or virtual. Some people will prefer to speak their ideas, dictating them. But spoken words are still likely to be turned into printed words (even if the print is simply on a display device), because reading is far faster and superior to listening. Reading can be done quickly: it is possible to read around three hundred words per minute and to skim, jumping ahead and back, effectively acquiring information at rates in the thousands of words per minute. Listening is slow and serial, usually at around sixty words per minute, and although this rate can be doubled or tripled with speech compression technologies and training, it is still slower than reading and not easy to skim. But the new media and new technologies will supplement the old, so that writing will no longer dominate as much as it did in the past, when it was the only medium widely available. Now that anyone can type and dictate, take photographs and videos, draw animated scenes, and creatively produce experiences that in the twentieth century required huge amounts of technology and large crews of specialized workers, the types of devices that allow us to do these tasks and the ways they are controlled will proliferate.

The role of writing in civilization has changed over its five thousand years of existence. Today, writing has become increasingly common, although increasingly as short, informal messages. We now communicate using a wide variety of media: voice, video, handwriting, and typing, sometimes with all ten fingers, sometimes just with the thumbs, and sometimes by gestures. Over time, the ways by which we interact and communicate change with technology. But because the fundamental psychology of human beings will remain unchanged, the design rules in this book will still apply.

Of course, it isn't just communication and writing that has changed. Technological change has impacted every sphere of our lives, from the way education is conducted, to medicine, foods, clothing, and transportation. We now can manufacture things at home, using 3-D printers. We can play games with partners around the world. Cars are capable of driving themselves, and their engines have changed from internal combustion to an assortment of pure electric and hybrids. Name an industry or an activity and if it hasn't already been transformed by new technologies, it will be.

Technology is a powerful driver for change. Sometimes for the better, sometimes for the worse. Sometimes to fulfill important needs, and sometimes simply because the technology makes the change possible.

How Long Does It Take to Introduce a New Product?

How long does it take for an idea to become a product? And after that, how long before the product becomes a long-lasting success? Inventors and founders of startup companies like to think the interval from idea to success is a single process, with the total measured in months. In fact, it is multiple processes, where the total time is measured in decades, sometimes centuries.

Technology changes rapidly, but people and culture change slowly. Change is, therefore, simultaneously rapid and slow. It can take months to go from invention to product, but then decades—sometimes many decades—for the product to get accepted. Older products linger on long after they should have become obsolete, long after they should have disappeared. Much of daily life is dictated by conventions that are centuries old, that no longer make any sense, and whose origins have been forgotten by all except the historian.

Even our most modern technologies follow this time cycle: fast to be invented, slow to be accepted, even slower to fade away and die. In the early 2000s, the commercial introduction of gestural control for cell phones, tablets, and computers radically transformed the way we interacted with our devices. Whereas all previous electronic devices had numerous knobs and buttons on the outside, physical keyboards, and ways of calling up numerous menus of commands, scrolling through them, and selecting the desired command, the new devices eliminated almost all physical controls and menus.

Was the development of tablets controlled by gestures revolutionary? To most people, yes, but not to technologists.

Touch-sensitive displays that could detect the positions of simultaneous finger presses (even if by multiple people) had been in the research laboratories for almost thirty years (these are called multitouch displays). The first devices were developed by the University of Toronto in the early 1980s. Mitsubishi developed a product that it sold to design schools and research laboratories, in which many of today's gestures and techniques were being explored. Why did it take so long for these multitouch devices to become successful products? Because it took decades to transform the research technology into components that were inexpensive and reliable enough for everyday products. Numerous small companies tried to manufacture screens, but the first devices that could handle multiple touches were either very expensive or unreliable.

There is another problem: the general conservatism of large companies. Most radical ideas fail: large companies are not tolerant of failure. Small companies can jump in with new, exciting ideas because if they fail, well, the cost is relatively low. In the world of high technology, many people get new ideas, gather together a few friends and early risk-seeking employees, and start a new company to exploit their visions. Most of these companies fail. Only a few will be successful, either by growing into a larger company or by being purchased by a large company.

You may be surprised by the large percentage of failures, but that is only because they are not publicized: we only hear about the tiny few that become successful. Most startup companies fail, but failure in the high-tech world of California is not considered bad. In fact, it is considered a badge of honor, for it means that the company saw a future potential, took the risk, and tried. Even though the company failed, the employees learned lessons that make their next attempt more likely to succeed. Failure can occur for many reasons: perhaps the marketplace is not ready; perhaps the technology is not ready for commercialization; perhaps the company runs out of money before it can gain traction.

When one early startup company, Fingerworks, was struggling to develop an affordable, reliable touch surface that distinguished among multiple fingers, it almost quit because it was about to run out of money. Apple however, anxious to get into this market, bought Fingerworks. When it became part of Apple, its financial needs were met and Fingerworks technology became the driving force behind Apple's new products. Today, devices controlled by gestures are everywhere, so this type of interaction seems natural and obvious, but at the time, it was neither natural nor obvious. It took almost three decades from the invention of multitouch before companies were able to manufacture the technology with the required robustness, versatility, and very low cost necessary for the idea to be deployed in the home consumer market. Ideas take a long time to traverse the distance from conception to successful product.

VIDEOPHONE: CONCEIVED IN 1879—STILL NOT HERE

The Wikipedia article on videophones, from which Figure 7.3 was taken, said: "George du Maurier's cartoon of 'an electric camera-obscura' is often cited as an early prediction of television and also anticipated the videophone, in wide screen formats and flat screens." Although the title of the drawing gives credit to Thomas Edison, he had nothing to do with this. This is sometimes called Stigler's law: the names of famous people often get attached to ideas even though they had nothing to do with them.

The world of product design offers many examples of Stigler's law. Products are thought to be the invention of the company that most successfully capitalized upon the idea, not the company that originated it. In the world of products, original ideas are the easy part. Actually producing the idea as a successful product is what is hard. Consider the idea of a video conversation. Thinking of the idea was so easy that, as we see in Figure 7.3, *Punch* magazine illustrator du Maurier could draw a picture of what it might look like only two years after the telephone was invented. The fact that he could do this probably meant that the idea was already circulating. By the late 1890s, Alexander Graham Bell had thought through a number of the design issues. But the wonderful scenario illustrated

FIGURE 7.3 **Predicting the Future: The Videophone in 1879.** The caption reads: "Edison's Telephonoscope (transmits light as well as sound). (*Every evening, before going to bed, Pater- and Materfamilias set up an electric camera-obscura over their bedroom mantel-piece, and gladden their eyes with the sight of their children at the Antipodes, and converse gaily with them through the wire.*") (Published in the December 9, 1878, issue of *Punch* magazine. From "Telephonoscope," Wikipedia.)

by du Maurier has still not become reality, one and one-half centuries later. Today, the videophone is barely getting established as a means of everyday communication.

It is extremely difficult to develop all the details required to ensure that a new idea works, to say nothing of finding components that can be manufactured in sufficient quantity, reliability, and affordability. With a brand-new concept, it can take decades before the public will endorse it. Inventors often believe their new ideas will revolutionize the world in months, but reality is harsher. Most new inventions fail, and even the few that succeed take decades to do so. Yes, even the ones we consider "fast." Most of the time, the technology is unnoticed by the public as it circulates around the research laboratories of the world or is tried by a few unsuccessful startup companies or adventurous early adopters.

Ideas that are too early often fail, even if eventually others introduce them successfully. I've seen this happen several times. When I first joined Apple, I watched as it released one of the very first commercial digital cameras: the Apple QuickTake. It failed. Probably you are unaware that Apple ever made cameras. It failed because the technology was limited, the price high, and the world simply wasn't ready to dismiss film and chemical processing of photographs. I was an adviser to a startup company that produced the world's first digital picture frame. It failed. Once again, the technology didn't quite support it and the product was relatively expensive. Obviously today, digital cameras and digital photo frames are extremely successful products, but neither Apple nor the startup I worked with are part of the story.

Even as digital cameras started to gain a foothold in photography, it took several decades before they displaced film for still photographs. It is taking even longer to replace film-based movies with those produced on digital cameras. As I write this, only a small number of films are made digitally, and only a small number of theaters project digitally. How long has the effort been going on? It is difficult to determine when the effort stated, but it has been a very long time. It took decades for high-definition television to replace the standard, very poor resolution of the previous generation (NTSC in the United States and PAL and SECAM elsewhere). Why so long to get to a far better picture, along with far better sound? People are very conservative. Broadcasting stations would have to replace all their equipment. Homeowners would need new sets. Overall, the only people who push for changes of this sort are the technology enthusiasts and the equipment manufacturers. A bitter fight between the television broadcasters and the computer industry, each of which wanted different standards, also delayed adoption (described in Chapter 6).

In the case of the videophone shown in Figure 7.3, the illustration is wonderful but the details are strangely lacking. Where would the video camera have to be located to display that wonderful panorama of the children playing? Notice that "Pater- and Materfamilias" are sitting in the dark (because the video image is projected by a "camera obscura," which has a very weak output). Where is the video camera that films the parents, and if they sit in the dark, how can they be visible? It is also interesting that although the video quality looks even better than we could achieve today, sound is still being picked up by trumpet-shaped telephones whose users need to hold the speaking tube to their face and talk (probably loudly). Thinking of the concept of a video connection was relatively easy. Thinking through the details has been very difficult, and then being able to build it and put it into practice—well, it is now considerably over a century since that picture was drawn and we are just barely able to fulfill that dream. Barely.

It took forty years for the first working videophones to be created (in the 1920s), then another ten years before the first product (in the mid-1930s, in Germany), which failed. The United States didn't try commercial videophone service until the 1960s, thirty years after Germany; that service also failed. All sorts of ideas have been tried including dedicated videophone instruments, devices using the home television set, video conferencing with home personal computers, special video-conferencing rooms in universities and companies, and small video telephones, some of which might be worn on the wrist. It took until the start of the twenty-first century for usage to pick up.

Video conferencing finally started to become common in the early 2010s. Extremely expensive videoconferencing suites have been set up in businesses and universities. The best commercial systems make it seem as if you are in the same room with the distant participants, using high-quality transmission of images and multiple, large monitors to display life-size images of people sitting across the table (one company, Cisco, even sells the table). This is 140 years from the first published conception, 90 years since the first practical demonstration, and 80 years since the first commercial release. Moreover, the cost, both for the equipment at each location and for the data-transmission charges, are much higher than the average person or business can afford: right now they are mostly used in corporate offices. Many people today do engage in videoconferencing from their smart display devices, but the experience is not nearly as good as provided by the best commercial facilities. Nobody would confuse these experiences with being in the same room as the participants, something that the highest-quality commercial facilities aspire to (with remarkable success).

Every modern innovation, especially the ones that significantly change lives, takes multiple decades to move from concept to company success A rule of thumb is twenty years from first demonstrations in research laboratories to commercial product, and then a decade or two from first commercial release to widespread adoption. Except that actually, most innovations fail completely and never reach the public. Even ideas that are excellent and will eventually succeed frequently fail when first introduced. I've been associated with a number of products that failed upon introduction, only to be very successful later when reintroduced (by other companies), the real difference being the timing. Products that failed at first commercial introduction include the first American automobile (Duryea), the first typewriters, the first digital cameras, and the first home computers (for example, the Altair 8800 computer of 1975).

THE LONG PROCESS OF DEVELOPMENT OF THE TYPEWRITER KEYBOARD

The typewriter is an ancient mechanical device, now found mostly in museums, although still in use in newly developing nations. In addition to having a fascinating history, it illustrates the difficulties of introducing new products into society, the influence of marketing upon design, and the long, difficult path leading to new product acceptance. The history affects all of us because the typewriter provided the world with the arrangement of keys on today's keyboards, despite the evidence that it is not the most efficient arrangement. Tradition and custom coupled with the large number of people already used to an existing scheme makes change difficult or even impossible. This is the legacy problem once again: the heavy momentum of legacy inhibits change.

Developing the first successful typewriter was a lot more than simply figuring out a reliable mechanism for imprinting the letters upon the paper, although that was a difficult task by itself. One question was the user interface: how should the letters be presented to the typist? In other words, the design of the keyboard.

Consider the typewriter keyboard, with its arbitrary, diagonally sloping arrangement of keys and its even more arbitrary arrangement of their letters. Christopher Latham Sholes designed the current standard keyboard in the 1870s. His typewriter design, with its weirdly organized keyboard, eventually became the Remington typewriter, the first successful typewriter: its keyboard layout was soon adopted by everyone.

The design of the keyboard has a long and peculiar history. Early typewriters experimented with a wide variety of layouts, using three basic themes. One was circular, with the letters laid out alphabetically; the operator would find the proper spot and depress a lever, lift a rod, or do whatever other mechanical operation the device required. Another popular layout was similar to a piano keyboard, with the letters laid out in a long row; some of the early keyboards, including an early version by Sholes, even had black and white keys. Both the circular layout and the piano keyboard proved awkward. In the end, the typewriter keyboards all ended up using multiple rows of keys in a rectangular configuration, with different companies using different arrangements of the letters. The levers manipulated by the keys were large and ungainly, and the size, spacing, and arrangement of the keys were dictated by these mechanical considerations, not by the characteristics of the human hand. Hence the keyboard sloped and the keys were laid out in a diagonal pattern to provide room for the mechanical linkages. Even though we no longer use mechanical linkages, the keyboard design is unchanged, even for the most modern electronic devices.

Alphabetical ordering of keys seems logical and sensible: Why did it change? The reason is rooted in the early technology of keyboards. Early typewriters had long levers attached to the keys. The levers moved individual typebars to contact the typing paper,

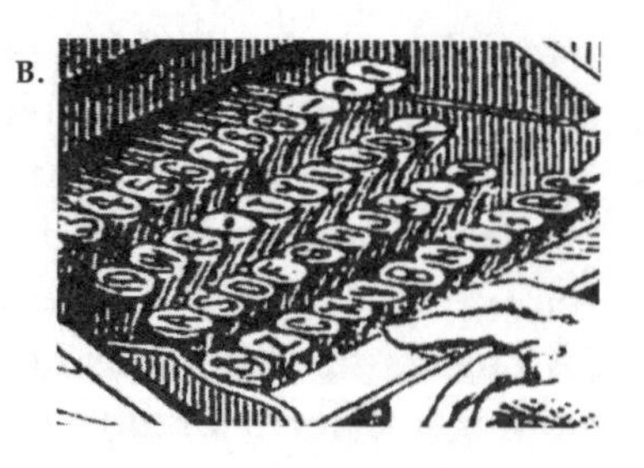

FIGURE 7.4. **The 1872 Sholes Typewriter.** Remington, the manufacturer of the first successful typewriter, also made sewing machines. Figure A shows the influence of the sewing machine upon the design with the use of a foot pedal for what eventually became the "return" key. A heavy weight hung from the frame advanced the carriage after each letter was struck, or when the large, rectangular plate under the typist's left hand was depressed (this is the "space bar"). Pressing the foot pedal raised the weight. Figure B shows a blowup of the keyboard. Note that the second row shows a period (.) instead of R. From *Scientific American's* "The Type Writer" (Anonymous, 1872).

usually from behind (the letters being typed could not be seen from the front of the typewriter). These long type arms would often collide and lock together, requiring the typist to separate them manually. To avoid the jamming, Sholes arranged the keys and the typebars so that letters that were frequently typed in sequence did not come from adjacent typebars. After a few iterations and experiments, a standard emerged, one that today governs keyboards used throughout the world, although with regional variations. The top row of the American keyboard has the keys Q W E R T Y U I O P, which gives rise to the name of this layout: QWERTY. The world has adopted the basic layout, although in Europe, for example, one can find QZERTY, AZERTY, and QWERTZ. Different languages use different alphabets, so obviously a number of keyboards had to move keys around to make room for additional characters.

Note that popular legend has it that the keys were placed so as to slow down the typing. This is wrong: the goal was to have the mechanical typebars approach one another at large angles, thus minimizing the chance of collision. In fact, we now know that the

QWERTY arrangement guarantees a fast typing speed. By placing letters that form frequent pairs relatively far apart, typing is speeded because it tends to make letter pairs be typed with different hands.

There is an unconfirmed story that a salesperson rearranged the keyboard to make it possible to type the word *typewriter* on the second row, a change that violated the design principle of separating letters that were typed sequentially. Figure 7.4B shows that the early Sholes keyboard was not QWERTY: the second row of keys had a period (.) where today we have R, and the P and R keys were on the bottom row (as well as other differences). Moving the R and P from the fourth row to the second makes it possible to type the word *typewriter* using only keys on the second row.

There is no way to confirm the validity of the story. Moreover, I have only heard it describe the interchange of the period and R keys, with no discussion of the P key. For the moment, suppose the story were true: I can imagine the engineering minds being outraged. This sounds like the traditional clash between the hard-headed, logical engineers and the noncomprehending sales and marketing force. Was the salesperson wrong? (Note that today we would call this a marketing decision, but the profession of marketing didn't exist yet.) Well, before taking sides, realize that until then, every typewriter company had failed. Remington was going to come out with a typewriter with a weird arrangement of the keys. The sales staff were right to be worried. They were right to try anything that might enhance the sales efforts. And indeed, they succeeded: Remington became the leader in typewriters. Actually, its first model did not succeed. It took quite a while for the public to accept the typewriter.

Was the keyboard really changed to allow the word *typewriter* to be typed on one row? I cannot find any solid evidence. But it is clear that the positions of R and P were moved to the second row: compare Figure 7.4B with today's keyboard.

The keyboard was designed through an evolutionary process, but the main driving forces were mechanical and marketing. Even though jamming isn't a possibility with electronic keyboards and computers and the style of typing has changed, we are committed to this keyboard, stuck with it forever. But don't despair: it really is a good arrangement. One legitimate area of concern is the high incidence of a kind of injury that befalls typists: carpal tunnel syndrome. This ailment is a result of frequent and prolonged repetitive motions of the hand and wrist, so it is common among typists, musicians, and people who do a lot of handwriting, sewing, some sports, and assembly line work. Gestural keyboards, such as the one shown in Figure 7.2D, might reduce the incidence. The US National Institute of Health advises, "Ergonomic aids, such as split keyboards, keyboard trays, typing pads, and wrist braces, may be used to improve wrist posture during typing. Take frequent breaks when typing and always stop if there is tingling or pain."

August Dvorak, an educational psychologist, painstakingly developed a better keyboard in the 1930s. The Dvorak keyboard layout is indeed superior to that of QWERTY, but not to the extent claimed. Studies in my laboratory showed that the typing speed on a QWERTY was only slightly slower than on a Dvorak, not different enough to make upsetting the legacy worthwhile. Millions of people would have to learn a new style of typing. Millions of typewriters would have to be changed. Once a standard is in place, the vested interests of existing practices impede change, even where the change would be an improvement. Moreover, in the case of QWERTY versus Dvorak, the gain is simply not worth the pain. "Good enough" triumphs again.

What about keyboards in alphabetical order? Now that we no longer have mechanical constraints on keyboard ordering, wouldn't they at least be easier to learn? Nope. Because the letters have to be laid out in several rows, just knowing the alphabet isn't enough. You also have to know where the rows break, and today, every alphabetic keyboard breaks the rows at different points. One great advantage of QWERTY—that frequent letter pairs are typed with opposite hands—would no longer be true. In other words, forget it. In my studies, QWERTY and Dvorak typing speeds were considerably faster than those on alphabetic keyboards. And an alphabetical arrangement of the keys was no faster than a random arrangement.

Could we do better if we could depress more than one finger at a time? Yes, court stenographers can out-type anyone else. They use chord keyboards, typing syllables, not individual letters, directly onto the page—each syllable represented by the simultaneous pressing of keys, each combination being called a "chord." The most common keyboard for American law court recorders requires between two and six keys to be pressed simultaneously to code the digits, punctuation, and phonetic sounds of English.

Although chord keyboards can be very fast—more than three hundred words per minute is common—the chords are difficult to learn and to retain; all the knowledge has to be in the head. Walk up to any regular keyboard and you can use it right away. Just search for the letter you want and push that key. With a chord keyboard, you have to press several keys simultaneously. There is no way to label the keys properly and no way to know what to do just by looking. The casual typist is out of luck.

Two Forms of Innovation: Incremental and Radical

There are two major forms of product innovation: one follows a natural, slow evolutionary process; the other is achieved through radical new development. In general, people tend to think of innovation as being radical, major changes, whereas the most common and powerful form of it is actually small and incremental.

Although each step of incremental evolution is modest, continual slow, steady improvements can result in rather significant changes over time. Consider the automobile. Steam-driven vehicles (the first automobiles) were developed in the late 1700s. The first commercial automobile was built in 1888 by the German Karl Benz (his company, Benz & Cie, later merged with Daimler and today is known as Mercedes-Benz).

Benz's automobile was a radical innovation. And although his firm survived, most of its rivals did not. The first American automobile company was Duryea, which only lasted a few years: being first does not guarantee success. Although the automobile itself was a radical innovation, since its introduction it has advanced through continual slow, steady improvement, year after year: over a century of incremental innovation (with a few radical changes in components). Because of the century of incremental enhancement, today's automobiles are much quieter, faster, more efficient, more comfortable, safer, and less expensive (adjusted for inflation) than those early vehicles.

Radical innovation changes paradigms. The typewriter was a radical innovation that had dramatic impact upon office and home writing. It helped provide a role for women in offices as typists and secretaries, which led to the redefinition of the job of secretary to be a dead end rather than the first step toward an executive position. Similarly, the automobile transformed home life, allowing people to live at a distance from their work and radically impacting the world of business. It also turned out to be a massive source of air pollution (although it did eliminate horse manure from city streets). It is a major cause of accidental death, with a worldwide fatality rate of over one million each year. The introduction of electric lighting, the airplane, radio, television, home computer, and social networks all had massive social impacts. Mobile phones changed the phone industry, and the use of the technical communication system called packet switching led to the Internet. These are radical innovations. Radical innovation changes lives and industries. Incremental innovation makes things better. We need both.

INCREMENTAL INNOVATION

Most design evolves through incremental innovation by means of continual testing and refinement. In the ideal case, the design is tested, problem areas are discovered and modified, and then the product is continually retested and remodified. If a change makes matters worse, well, it just gets changed again on the next go-round. Eventually the bad features are modified into good ones, while the good ones are kept. The technical term for this process is *hill climbing,* analogous to climbing a hill blindfolded. Move your foot in one direction. If it is downhill, try another direction. If the direction is uphill, take one step. Keep doing this until you have reached a point where all steps would be downhill; then you are at the top of the hill, or at least at a local peak.

Hill climbing. This method is the secret to incremental innovation. This is at the heart of the human-centered design process discussed in Chapter 6. Does hill climbing always work? Although it guarantees that the design will reach the top of the hill, what if the design is not on the best possible hill? Hill climbing cannot find higher hills: it can only find the peak of the hill it started from. Want to try a different hill? Try radical innovation, although that is as likely to find a worse hill as a better one.

RADICAL INNOVATION

Incremental innovation starts with existing products and makes them better. Radical innovation starts fresh, often driven by new technologies that make possible new capabilities. Thus, the invention of vacuum tubes was a radical innovation, paving the way for rapid advances in radio and television. Similarly, the invention of the transistor allowed dramatic advances in electronic devices, computational power, increased reliability, and lower costs. The development of GPS satellites unleashed a torrent of location-based services.

A second factor is the reconsideration of the meaning of technology. Modern data networks serve as an example. Newspapers, magazines, and books were once thought of as part of the publishing industry, very different from radio and television broadcasting. All of these were different from movies and music. But once the Internet took hold, along with enhanced and inexpensive computer power and displays, it became clear that all of these disparate industries were really just different forms of information providers, so that all could be conveyed to customers by a single medium. This redefinition collapses together the publishing, telephone, television and cable broadcasting, and music industries. We still have books, newspapers, and magazines, television shows and movies, musicians and music, but the way by which they are distributed has changed, thereby requiring massive restructuring of their corresponding industries. Electronic games, another radical innovation, are combining with film and video on the one hand, and books on the other, to form new types of interactive engagement. The collapsing of industries is still taking place, and what will replace them is not yet clear.

Radical innovation is what many people seek, for it is the big, spectacular form of change. But most radical ideas fail, and even those that succeed can take decades and, as this chapter has already illustrated, they may take centuries to succeed. Incremental product innovation is difficult, but these difficulties pale to insignificance compared to the challenges faced by radical innovation. Incremental innovations occur by the millions each year; radical innovation is far less frequent.

What industries are ready for radical innovation? Try education, transportation, medicine, and housing, all of which are overdue for major transformation.

The Design of Everyday Things: 1988–2038

Technology changes rapidly, people and culture change slowly. Or as the French put it:

> *Plus ça change, plus c'est la même chose.*
> *The more things change, the more they are the same.*

Evolutionary change to people is always taking place, but the pace of human evolutionary change is measured in thousands of years. Human cultures change somewhat more rapidly over periods measured in decades or centuries. Microcultures, such as the way by which teenagers differ from adults, can change in a generation. What this means is that although technology is continually introducing new means of doing things, people are resistant to changes in the way they do things.

Consider three simple examples: social interaction, communication, and music. These represent three different human activities, but each is so fundamental to human life that all three have persisted throughout recorded history and will persist, despite major changes in the technologies that support these activities. They are akin to eating: new technologies will change the types of food we eat and the way it is prepared, but will never eliminate the need to eat. People often ask me to predict "the next great change." My answer is to tell them to examine some fundamentals, such as social interaction, communication, sports and play, music and entertainment. The changes will take place within spheres of activity such as these. Are these the only fundamentals? Of course not: add education (and learning), business (and commerce), transportation, self-expression, the arts, and of course, sex. And don't forget important sustaining activities, such as the need for good health, food and drink, clothing, and housing. Fundamental needs will also stay the same, even if they get satisfied in radically different ways.

The Design of Everyday Things was first published in 1988 (when it was called *The Psychology of Everyday Things*). Since the original publication, technology has changed so much that even though the principles remained constant, many of the examples from 1988 are no longer relevant. The technology of interaction has changed. Oh yes, doors and switches, faucets and taps still provide the same difficulties they did back then, but now we have new sources of difficulties and confusion. The same principles that worked before still apply, but this time they must also be applied to intelligent machines, to the continuous interaction with large data sources, to social networks and to communication systems and products that enable lifelong interaction with friends and acquaintances across the world.

We gesture and dance to interact with our devices, and in turn they interact with us via sound and touch, and through multiple displays of all sizes—some that we wear; some on the floor, walls, or ceilings; and some projected directly into our eyes. We speak to our devices and they speak back. And as they get more and more intelligent, they take over many of the activities we thought that only people could do. Artificial intelligence pervades our lives and devices, from our thermostats to our automobiles. Technologies are always undergoing change.

AS TECHNOLOGIES CHANGE WILL PEOPLE STAY THE SAME?

As we develop new forms of interaction and communication, what new principles are required? What happens when we wear augmented reality glasses or embed more and more technology within our bodies? Gestures and body movements are fun, but not very precise.

For many millennia, even though technology has undergone radical change, people have remained the same. Will this hold true in the future? What happens as we add more and more enhancements inside the human body? People with prosthetic limbs will be faster, stronger, and better runners or sports players than normal players. Implanted hearing devices and artificial lenses and corneas are already in use. Implanted memory and communication devices will mean that some people will have permanently enhanced reality, never lacking for information. Implanted computational devices could enhance thinking, problem-solving, and decision-making. People might become cyborgs: part biology, part artificial technology. In turn, machines will become more like people, with neural-like computational abilities and humanlike behavior. Moreover, new developments in biology might add to the list of artificial supplements, with genetic modification of people and biological processors and devices for machines.

All of these changes raise considerable ethical issues. The long-held view that even as technology changes, people remain the same may no longer hold. Moreover, a new species is arising, artificial devices that have many of the capabilities of animals and people, sometimes superior abilities. (That machines might be better than people at some things has long been true: they are clearly stronger and faster. Even the simple desk calculator can do arithmetic better than we can, which is why we use them. Many computer programs can do advanced mathematics better than we can, which makes them valuable assistants.) People are changing; machines are changing. This also means that cultures are changing.

There is no question that human culture has been vastly impacted by the advent of technology. Our lives, our family size and living arrangements, and the role played by business and education in our lives are all governed by the technologies of the era. Modern communication technology changes the nature of joint work. As some people get advanced cognitive skills due to implants, while some machines gain enhanced human-qualities through advanced technologies, artificial intelligence, and perhaps bionic technologies, we can expect even more changes. Technology, people, and cultures: all will change.

THINGS THAT MAKE US SMART

Couple the use of full-body motion and gestures with high-quality auditory and visual displays that can be superimposed over the sounds and sights of the world to amplify them, to explain and annotate them, and we give to people power that exceeds anything ever known before. What do the limits of human memory mean when a machine can remind us of all that has happened before, at precisely the exact time the information is needed? One argument is that technology makes us smart: we remember far more than ever before and our cognitive abilities are much enhanced.

Another argument is that technology makes us stupid. Sure, we look smart with the technology, but take it away and we are worse off than before it existed. We have become dependent upon our technologies to navigate the world, to hold intelligent conversation, to write intelligently, and to remember.

Once technology can do our arithmetic, can remember for us, and can tell us how to behave, then we have no need to learn these things. But the instant the technology goes away, we are left helpless, unable to do any basic functions. We are now so dependent upon technology that when we are deprived, we suffer. We are unable to make our own clothes from plants and animal skins, unable to grow and harvest crops or catch animals. Without technology, we would starve or freeze to death. Without cognitive technologies, will we fall into an equivalent state of ignorance?

These fears have long been with us. In ancient Greece, Plato tells us that Socrates complained about the impact of books, arguing that reliance on written material would diminish not only memory but the very need to think, to debate, to learn through discussion. After all, said Socrates, when a person tells you something, you can question the statement, discuss and debate it, thereby enhancing the material and the understanding. With a book, well, what can you do? You can't argue back.

But over the years, the human brain has remained much the same. Human intelligence has certainly not diminished. True, we no longer learn how to memorize vast amounts of material. We no longer need to be completely proficient at arithmetic, for calculators—present as dedicated devices or on almost every computer or phone—take care of that task for us. But does that make us stupid? Does the fact that I can no longer remember my own phone number indicate my growing feebleness? No, on the contrary, it unleashes the mind from the petty tyranny of tending to the trivial and allows it to concentrate on the important and the critical.

Reliance on technology is a benefit to humanity. With technology, the brain gets neither better nor worse. Instead, it is the task that changes. Human plus machine is more powerful than either human or machine alone.

The best chess-playing machine can beat the best human chess player. But guess what, the combination of human plus machine can beat the best human and the best machine. Moreover, this winning combination need not have the best human or machine. As MIT professor Erik Brynjolfsson explained at a meeting of the National Academy of Engineering:

> *The best chess player in the world today is not a computer or a human but a team of humans and computers working together. In freestyle chess competitions, where teams of humans and computers compete,*

> *the winners tend not to be the teams with the most powerful computers or the best chess players. The winning teams are able to leverage the unique skills of humans and computers to work together. That is a metaphor for what we can do going forward: have people and technology work together in new ways to create value.* (Brynjolfsson, 2012.)

Why is this? Brynjolfsson and Andrew McAfee quote the world-champion human chess player Gary Kasparov, explaining why "the overall winner in a recent freestyle tournament had neither the best human players nor the most powerful computers." Kasparov described a team consisting of:

> *a pair of amateur American chess players using three computers at the same time. Their skill at manipulating and "coaching" their computers to look very deeply into positions effectively counteracted the superior chess understanding of their grandmaster opponents and the greater computational power of other participants. Weak human + machine + better process was superior to a strong computer alone and, more remarkably, superior to a strong human + machine + inferior process.* (Brynjolfsson & McAfee, 2011.)

Moreover, Brynjolfsson and McAfee argue that the same pattern is found in many activities, including both business and science: "The key to winning the race is not to compete against machines but to compete with machines. Fortunately, humans are strongest exactly where computers are weak, creating a potentially beautiful partnership."

The cognitive scientist (and anthropologist) Edwin Hutchins of the University of California, San Diego, has championed the power of distributed cognition, whereby some components are done by people (who may be distributed across time and space); other components, by our technologies. It was he who taught me how powerful this combination makes us. This provides the answer to the question: Does the new technology make us stupid? No, on the contrary, it changes the tasks we do. Just as the best chess player is a combination of human and technology, we, in combination with technology, are smarter than ever before. As I put it in my book *Things That Make Us Smart*, the power of the unaided mind is highly overrated. It is things that make us smart.

> *The power of the unaided mind is highly overrated. Without external aids, deep, sustained reasoning is difficult. Unaided memory, thought, and reasoning are all limited in power. Human intelligence is highly flexible and adaptive, superb at inventing procedures and objects that overcome its own limits. The real powers come from devising external aids that enhance cognitive abilities. How have we increased memory, thought and reasoning? By the invention of external aids: it is things that make us smart. Some assistance comes through cooperative, social behavior: some arises through exploitation of the information present in the environment; and some comes through the development of tools of thought—cognitive artifacts—that complement abilities and strengthen mental powers.* (The opening paragraph of Chapter 3, *Things That Make Us Smart*, 1993.)

The Future of Books

It is one thing to have tools that aid in writing conventional books, but quite another when we have tools that dramatically transform the book.

Why should a book comprise words and some illustrations meant to be read linearly from front to back? Why shouldn't it be composed of small sections, readable in whatever order is desired? Why shouldn't it be dynamic, with video and audio segments, perhaps changing according to who is reading it, including notes made by other readers or viewers, or incorporating the author's latest thoughts, perhaps changing even as it is being read, where the word *text* could mean anything: voice, video, images, diagrams, and words?

Some authors, especially of fiction, might still prefer the linear telling of tales, for authors are storytellers, and in stories, the order in which characters and events are introduced is important to build the suspense, keep the reader enthralled, and manage the emotional highs and lows that characterize great storytelling. But for nonfiction, for books like this one, order is not as important. This book does not attempt to manipulate your emotions, to keep you in suspense, or to have dramatic peaks. You should be able to experience it in the order you prefer, reading items out of sequence and skipping whatever is not relevant to your needs.

Suppose this book were interactive? If you have trouble understanding something, suppose you could click on the page and I would pop up and explain something. I tried that many years ago with three of my books, all combined into one interactive electronic book. But the attempt fell prey to the demons of product design: good ideas that appear too early will fail.

It took a lot of effort to produce that book. I worked with a large team of people from Voyager Books, flying to Santa Monica, California, for roughly a year of visits to film the excerpts and record my part. Robert Stein, the head of Voyager, assembled a talented team of editors, producers, videographers, interactive designers, and illustrators. Alas, the result was produced in a computer system called HyperCard, a clever tool developed by Apple but never really given full support. Eventually, Apple stopped supporting it and today, even though I still have copies of the original disks, they will not run on any existing machine. (And even if they could, the video resolution is very poor by today's standards.)

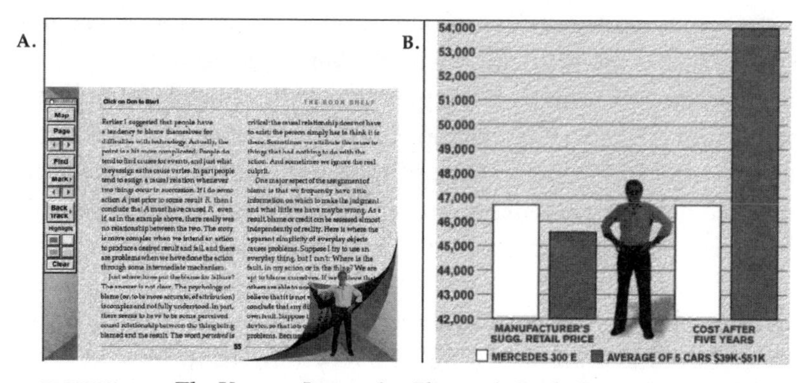

FIGURE 7.5. The Voyager Interactive Electronic Book. Figure A, on the left, is me stepping on to a page of *The Design of Everyday Things*. Figure B, on the right, shows me explaining a point about graph design in my book *Things That Make Us Smart*.

Notice the phrase "it took a lot of effort to produce that book." I don't even remember how many people were involved, but the credits include the following: editor-producer, art director–graphic designer, programmer, interface designers (four people, including me), the production team (twenty-seven people), and then special thanks to seventeen people.

Yes, today anybody can record a voice or video essay. Anyone can shoot a video and do simple editing. But to produce a professional-level multimedia book of roughly three hundred pages or two hours of video (or some combination) that will be read and enjoyed by people across the world requires an immense amount of talent and a variety of skills. Amateurs can do a five- or ten-minute video, but anything beyond that requires superb editing skills. Moreover, there has to be a writer, a cameraperson, a recording person, and a lighting person. There has to be a director to coordinate these activities and to select the best approach to each scene (chapter). A skilled editor is required to piece the segments together. An electronic book on the environment, Al Gore's interactive media book *Our Choice* (2011), lists a large number of job titles for the people responsible for this one book: publishers (two people), editor, production director, production editor, and production supervisor, software architect, user interface engineer, engineer, interactive graphics, animations, graphics design, photo editor, video editors (two), videographer, music, and cover designer. What is the future of the book? Very expensive.

The advent of new technologies is making books, interactive media, and all sorts of educational and recreational material more effective and pleasurable. Each of the many tools makes creation easier. As a result, we will see a proliferation of materials. Most will be amateurish, incomplete, and somewhat incoherent. But even amateur productions can serve valuable functions in our lives, as the immense proliferation of homemade videos available on the Internet demonstrate, teaching us everything from how to cook Korean *pajeon*, repair a faucet, or understand Maxwell's equations of electromagnetic waves. But for high-quality professional material that tells a coherent story in a way that is reliable, where the facts have been checked and the message authoritative, where the material will flow, experts are needed. The mix of technologies and tools makes quick and rough creation easier, but polished and professional level material much more difficult. The society of the future: something to look forward to with pleasure, contemplation, and dread.

The Moral Obligations of Design

That design affects society is hardly news to designers. Many take the implications of their work seriously. But the conscious manipulation of society has severe drawbacks, not the least of which is the fact that not everyone agrees on the appropriate goals. Design, therefore, takes on political significance; indeed, design philosophies vary in important ways across political systems. In Western cultures, design has reflected the capitalistic importance of the marketplace, with an emphasis on exterior features deemed to be attractive to the purchaser. In the consumer economy, taste is not the criterion in the marketing of expensive foods or drinks, usability is not the primary criterion in the marketing of home and office appliances. We are surrounded with objects of desire, not objects of use.

NEEDLESS FEATURES, NEEDLESS MODELS: GOOD FOR BUSINESS, BAD FOR THE ENVIRONMENT

In the world of consumable products, such as food and news, there is always a need for more food and news. When the product is consumed, then the customers are consumers. A never-ending cycle. In the world of services, the same applies. Someone has to cook and serve the food in a restaurant, take care of us when we are sick, do the daily transactions we all need. Services can be self-sustaining because the need is always there.

But a business that makes and sells durable goods faces a problem: As soon as everyone who wants the product has it, then there is no need for more. Sales will cease. The company will go out of business.

In the 1920s, manufacturers deliberately planned ways of making their products become obsolete (although the practice had existed long before then). Products were built with a limited life span. Automobiles were designed to fall apart. A story tells of Henry Ford's buying scrapped Ford cars and having his engineers disassemble them to see which parts failed and which were still in good shape. Engineers assumed this was done to find the weak parts and make them stronger. Nope. Ford explained that he wanted to find the parts that were still in good shape. The company could save money if they redesigned these parts to fail at the same time as the others.

Making things fail is not the only way to sustain sales. The women's clothing industry is an example: what is fashionable this year is not next year, so women are encouraged to replace their wardrobe every season, every year. The same philosophy was soon extended to the automobile industry, where dramatic style changes on a regular basis made it obvious which people were up to date; which people were laggards, driving old-fashioned vehicles. The same is true for our smart screens, cameras, and TV sets. Even the kitchen and laundry, where appliances used to last for decades, have seen the impact of fashion. Now, out-of-date features, out-of-date styling, and even out-of-date colors entice homeowners to change. There are some gender differences. Men are not as sensitive as women to fashion in clothes, but they more than make up for the difference by their interest in the latest fashions in automobiles and other technologies.

But why purchase a new computer when the old one is functioning perfectly well? Why buy a new cooktop or refrigerator, a new phone or camera? Do we really need the ice cube dispenser in the door of the refrigerator, the display screen on the oven door, the navigation system that uses three-dimensional images? What is the cost to the environment for all the materials and energy used to manufacture the new products, to say nothing of the problems of disposing safely of the old?

Another model for sustainability is the subscription model. Do you have an electronic reading device, or music or video player? Subscribe to the service that provides articles and news, music and entertainment, video and movies. These are all consumables, so even though the smart screen is a fixed, durable good, the subscription guarantees a steady stream of money in return for services. Of course this only works if the manufacturer of the durable good is also the provider of services. If not, what alternatives are there?

Ah, the model year: each year a new model can be introduced, just as good as the previous year's model, only claiming to be better. It always increases in power and features. Look at all the new features. How did you ever exist without them? Meanwhile, scientists, engineers, and inventors are busy developing yet newer technologies. Do you like your television? What if it were in three dimensions? With multiple channels of surround sound? With virtual goggles so you are surrounded by the images, 360 degrees' worth? Turn your head or body and see what is happening behind you. When you watch sports, you can be inside the team, experiencing the game the way the team does. Cars not only will drive themselves to make you safer, but provide lots of entertainment along the way. Video games will keep adding layers and chapters, new story lines and characters, and of course, 3-D virtual environments. Household appliances will talk to one another, telling remote households the secrets of our usage patterns.

The design of everyday things is in great danger of becoming the design of superfluous, overloaded, unnecessary things.

Design Thinking and Thinking About Design

Design is successful only if the final product is successful—if people buy it, use it, and enjoy it, thus spreading the word. A design that people do not purchase is a failed design, no matter how great the design team might consider it.

Designers need to make things that satisfy people's needs, in terms of function, in terms of being understandable and usable, and in terms of their ability to deliver emotional satisfaction, pride, and delight. In other words, the design must be thought of as a total experience.

But successful products need more than a great design. They have to be able to be produced reliably, efficiently, and on schedule. If the design complicates the engineering requirements so much that they cannot be realized within the cost and scheduling constraints, then the design is flawed. Similarly, if manufacturing cannot produce the product, then the design is flawed.

Marketing considerations are important. Designers want to satisfy people's needs. Marketing wants to ensure that people actually buy and use the product. These are two different sets of requirements: design must satisfy both. It doesn't matter how great the design is if people don't buy it. And it doesn't matter how many people buy something if they are going to dislike it when they start using it. Designers will be more effective as they learn more about sales and marketing, and the financial parts of the business.

Finally, products have a complex life cycle. Many people will need assistance in using a device, either because the design or the manual is not clear, or because they are doing something novel that was not considered in the product development, or for numerous other reasons. If the service provided to these people is inadequate, the product will suffer. Similarly if the device must be maintained, repaired, or upgraded, how this is managed affects people's appreciation of the product.

In today's environmentally sensitive world, the full life cycle of the product must be taken into consideration. What are the environmental costs of the materials, of the manufacturing process, of distribution, servicing, and repairs? When it is time to replace the unit, what is the environmental impact of recycling or otherwise reusing the old?

The product development process is complex and difficult. But to me, that is why it can be so rewarding. Great products pass through a gauntlet of challenges. To satisfy the myriad needs requires skill as well as patience. It requires a combination of high technical skills, great business skills, and a large amount of personal social skills for interacting with the many other groups that are involved, all of whom have their own agendas, all of which believe their requirements to be critical.

Design consists of a series of wonderful, exciting challenges, with each challenge being an opportunity. Like all great drama, it has its emotional highs and lows, peaks and valleys. The great products overcome the lows and end up high.

Now you are on your own. If you are a designer, help fight the battle for usability. If you are a user, then join your voice with those who cry for usable products. Write to manufacturers. Boycott unusable designs. Support good designs by purchasing them, even if it means going out of your way, even if it means spending a bit more. And voice your concerns to the stores that carry the products; manufacturers listen to their customers.

When you visit museums of science and technology, ask questions if you have trouble understanding. Provide feedback about the exhibits and whether they work well or poorly. Encourage museums to move toward better usability and understandability.

And enjoy yourself. Walk around the world examining the details of design. Learn how to observe. Take pride in the little things that help: think kindly of the person who so thoughtfully put them in. Realize that even details matter, that the designer may have had to fight to include something helpful. If you have difficulties, remember, it's not your fault: it's bad design. Give prizes to those who practice good design: send flowers. Jeer those who don't: send weeds.

Technology continually changes. Much is for the good. Much is not. All technology can be used in ways never intended by the inventors. One exciting development is what I call "the rise of the small."

THE RISE OF THE SMALL

I dream of the power of individuals, whether alone or in small groups, to unleash their creative spirits, their imagination, and their talents to develop a wide range of innovation. New technologies promise to make this possible. Now, for the first time in history, individuals can share their ideas, their thoughts and dreams. They can produce their own products, their own services, and make these available to anyone in the world. All can be their own master, exercising whatever special talents and interests they may have.

What drives this dream? The rise of small, efficient tools that empower individuals. The list is large and growing continuously. Consider the rise of musical explorations through conventional, electronic, and virtual instruments. Consider the rise of self-publishing, bypassing conventional publishers, printers and distributors, and replacing these with inexpensive electronic editions available to anyone in the world to download to e-book readers.

Witness the rise of billions of small videos, available to all. Some are simply self-serving, some are incredibly educational, and some are humorous, some serious. They cover everything from how to make spätzle to how to understand mathematics, or simply how to dance or play a musical instrument. Some films are purely for entertainment. Universities are getting into the act, sharing whole curricula, including videos of lectures. College students post their class assignments as videos and text, allowing the whole world to benefit from their efforts. Consider the same phenomenon in writing, reporting events, and the creation of music and art.

Add to these capabilities the ready availability of inexpensive motors, sensors, computation, and communication. Now consider the potential when 3-D printers increase in performance while decreasing in price, allowing individuals to manufacture custom items whenever they are required. Designers all over the world will publish their ideas and plans, enabling entire new industries of custom mass production. Small quantities can be made as inexpensively as large, and individuals might design their own items or rely on an ever-increasing number of freelance designers who will publish plans that can then be customized and printed at local 3-D print shops or within their own homes.

Consider the rise of specialists to help plan meals and cook them, to modify designs to fit needs and circumstances, to tutor on a wide variety of topics. Experts share their knowledge on blogs and on Wikipedia, all out of altruism, being rewarded by the thanks of their readers.

I dream of a renaissance of talent, where people are empowered to create, to use their skills and talents. Some may wish for the safety and security of working for organizations. Some may wish to start new enterprises. Some may do this as hobbies. Some may band together into small groups and cooperatives, the better to assemble the variety of skills required by modern technology, to help share their knowledge, to teach one another, and to assemble the critical mass that will always be needed, even for small projects. Some may hire themselves out to provide the necessary skills required of large projects, while still keeping their own freedom and authority.

In the past, innovation happened in the industrialized nations and with time, each innovation became more powerful, more complex, often bloated with features. Older technology was given to the developing nations. The cost to the environment was seldom considered. But with the rise of the small, with new, flexible, inexpensive technologies, the power is shifting. Today, anyone in the world can create, design, and manufacture. The newly developed nations are taking advantage, designing and building by themselves, for themselves. Moreover, out of necessity they develop advanced devices that require less power, that are simpler to make, maintain, and use. They develop medical procedures that don't require refrigeration or continual access to electric power. Instead of using handed-down technology, their results add value for all of us—call it handed-up technology.

With the rise of global interconnection, global communication, powerful design, and manufacturing methods that can be used by all, the world is rapidly changing. Design is a powerful equalizing tool: all that is needed is observation, creativity, and hard work—anyone can do it. With open-source software, inexpensive open-source 3-D printers, and even open-source education, we can transform the world.

AS THE WORLD CHANGES, WHAT STAYS THE SAME?

With massive change, a number of fundamental principles stay the same. Human beings have always been social beings. Social interaction and the ability to keep in touch with people across the world, across time, will stay with us. The design principles of this book will not change, for the principles of discoverability, of feedback, and of the power of affordances and signifiers, mapping, and conceptual models will always hold. Even fully autonomous, automatic machines will follow these principles for their interactions. Our technologies may change, but the fundamental principles of interaction are permanent.

ACKNOWLEDGMENTS

The original edition of this book was entitled *The Psychology of Everyday Things* (POET). This title is a good example of the difference between academics and industry. POET was a clever, cute title, much loved by my academic friends. When Doubleday/Currency approached me about publishing the paperback version of this book, the editors also said, "But of course, the title will have to be changed." Title changed? I was horrified. But I decided to follow my own advice and do some research on readers. I discovered that while the academic community liked the title and its cleverness, the business community did not. In fact, business often ignored the book because the title sent the wrong message. Bookstores placed the book in their psychology section (along with books on sex, love, and self-help). The final nail in the title's coffin came when I was asked to talk to a group of senior executives of a leading manufacturing company. The person who introduced me to the audience praised the book, damned the title, and asked his colleagues to read the book despite the title.

ACKNOWLEDGMENTS FOR *POET: PSYCHOLOGY OF EVERYDAY THINGS*

The book was conceived and the first few drafts written in the late 1980s while I was at the Applied Psychology Unit (the APU) in

Cambridge, England, a laboratory of the British Medical Research Council (the laboratory no longer exists). At the APU, I met another visiting American professor, David Rubin of Duke University, who was analyzing the recall of epic poetry. Rubin showed me that it wasn't all in memory: much of the information was in the world, or at least in the structure of the tale, the poetics, and the lifestyles of the people.

After spending the fall and winter in Cambridge, England, at the APU, I went to Austin, Texas, for the spring and summer (yes, the opposite order from what would be predicted by thinking of the weather at these two places). In Austin, I was at the Microelectronics and Computer Consortium (MCC), where I completed the manuscript. Finally, when I returned to my home base at the University of California, San Diego (UCSD), I revised the book several more times. I used it in classes and sent copies to a variety of colleagues for suggestions. I benefited greatly from my interactions at all these places: APU, MCC, and, of course, UCSD. The comments of my students and readers were invaluable, causing radical revision from the original structure.

My hosts at the APU in Britain were most gracious, especially Alan Baddeley, Phil Barnard, Thomas Green, Phil Johnson-Laird, Tony Marcel, Karalyn and Roy Patterson, Tim Shallice, and Richard Young. Peter Cook, Jonathan Grudin, and Dave Wroblewski were extremely helpful during my stay at the MCC in Texas (another institution that no longer exists). At UCSD, I especially wish to thank the students in Psychology 135 and 205: my undergraduate and graduate courses at UCSD entitled "Cognitive Engineering."

My understanding of how we interact with the world was developed and strengthened by years of debate and interaction with a very powerful team of people at UCSD from the departments of cognitive science, psychology, anthropology, and sociology, organized by Mike Cole, who met informally once a week for several years. The primary members were Roy d'Andrade, Aaron Cicourel, Mike Cole, Bud Mehan, George Mandler, Jean Mandler, Dave Rumelhart, and me. In later years, I benefited immensely from my interactions with Jim Hollan, Edwin Hutchins, and David Kirsh, all faculty members in the department of cognitive science at UCSD.

The early manuscript for POET was dramatically enhanced by critical readings by my colleagues: In particular, I am indebted to my editor at Basic Books, Judy Greissman, who provided patient critique through several revisions of POET.

My colleagues in the design community were most helpful with their comments: Mike King, Mihai Nadin, Dan Rosenberg, and Bill Verplank. Special thanks must be given to Phil Agre, Sherman De-Forest, and Jef Raskin, all of whom read the manuscript with care and provided numerous and valuable suggestions. Collecting the illustrations became part of the fun as I traveled the world with camera in hand. Eileen Conway and Michael Norman helped collect and organize the figures and illustrations. Julie Norman helped as she does on all my books, proofing, editing, commenting, and encouraging. Eric Norman provided valuable advice, support, and photogenic feet and hands.

Finally, my colleagues at the Institute for Cognitive Science at UCSD helped throughout—in part through the wizardry of international computer mail, in part through their personal assistance to the details of the process. I single out Bill Gaver, Mike Mozer, and Dave Owen for their detailed comments, but many helped out at one time or another during the research that preceded the book and the several years of writing.

ACKNOWLEDGMENTS FOR
***DESIGN OF EVERYDAY THINGS*, REVISED EDITION**

Because this new edition follows the organization and principles of the first, all the help given to me for that earlier edition applies to this one as well.

I have learned a lot in the years that have passed since the first edition of this book. For one thing, then I was an academic scholar. In the interim I have worked in several different companies. The most important experience was at Apple, where I began to appreciate how issues—budget, schedule, competitive forces, and the established base of products—that seldom concern scientists can dominate decisions in the world of business. While I was at Apple it had lost its way, but nothing is a better learning experience than a company in trouble: you have to be a fast learner.

I learned about schedules and budgets, about the competing demand of the different divisions, about the role of marketing, industrial design, and graphical, usability, and interactive design (today lumped together under the rubric of experience design). I visited numerous companies across the United States, Europe, and Asia and talked with numerous partners and customers. It was a great learning experience. I am indebted to Dave Nagel, who hired and then promoted me to vice president of advanced technology, and to John Scully, the first CEO I worked with at Apple: John had the correct vision of the future. I learned from many people, far too many to name (a quick review of the Apple people I worked closely with and who are still in my contact list reveals 240 names).

I learned about industrial design first from Bob Brunner, then from Jonathan (Joni) Ive. (Joni and I had to fight together to convince Apple management to produce his ideas. My, how Apple has changed!) Joy Mountford ran the design team in advanced technology and Paulien Strijland ran the usability testing group in the product division. Tom Erickson, Harry Saddler, and Austin Henderson worked for me in the User Experience Architect's office. Of particular significance to my increased understanding were Larry Tesler, Ike Nassi, Doug Solomon, Michael Mace, Rick LaFaivre, Guerrino De Luca, and Hugh Dubberly. Of special importance were the Apple Fellows Alan Kay, Guy Kawasaki, and Gary Starkweather. (I was originally hired as an Apple Fellow. All Fellows reported to the VP of advanced technology.) Steve Wozniak, by a peculiar quirk, was an Apple employee with me as his boss, which allowed me to spend a delightful afternoon with him. I apologize to those of you who were so helpful, but who I have not included here.

I thank my wife and critical reader, Julie Norman, for her patience in repeated careful readings of the manuscripts, telling me when I was stupid, redundant, and overly wordy. Eric Norman showed up as a young child in two of the photos of the first edition, and now, twenty-five years later, read the entire manuscript and provided cogent, valuable critiques. My assistant, Mimi Gardner, held off the e-mail onslaught, allowing me to concentrate upon writing, and of course my friends at the Nielsen Norman group provided inspiration. Thank you, Jakob.

Danny Bobrow of the Palo Alto Research Center, a frequent collaborator and coauthor of science papers for four decades, has provided continual advice and cogent critiques of my ideas. Lera Boroditsky shared her research on space and time with me, and further delighted me by leaving Stanford to take a job at the department I had founded, Cognitive Science, at UCSD.

I am of course indebted to Professor Yutaka Sayeki of the University of Tokyo for permission to use his story of how he managed the turn signals on his motorcycle. I used the story in the first edition, but disguised the name. A diligent Japanese reader figured out who it must have been, so for this edition, I asked Sayeki for permission to name him.

Professor Kun-Pyo Lee invited me to spend two months a year for three years at the Korea Advanced Institute for Science and Technology (KAIST) in its Industrial Design department, which gave me a much deeper insight into the teaching of design, Korean technology, and the culture of Northeast Asia, plus many new friends and a permanent love for kimchi.

Alex Kotlov, watching over the entrance to the building on Market Street in San Francisco where I photographed the destination control elevators, not only allowed me to photograph them, but then turned out to have read DOET!

In the years since publication of POET/DOET, I have learned a considerable amount about the practice of design. At IDEO I am indebted to David Kelly and Tim Brown, as well as fellow IDEO Fellows Barry Katz and Kristian Simsarian. I've had many fruitful discussions with Ken Friedman, former dean of the faculty of design at Swinburne University of Technology, Melbourne, as well as with my colleagues at many of the major schools of design around the world, in the United States, London, Delft, Eindhoven, Ivrea, Milan, Copenhagen, and Hong Kong.

And thanks to Sandra Dijkstra, my literary agent for almost thirty years, with POET being one of her first books, but who now has a large team of people and successful authors. Thanks, Sandy.

Andrew Haskin and Kelly Fadem, at the time students at CCA, the California College of the Arts in San Francisco, did all of the drawings in the book—a vast improvement over the ones in the first edition that I did myself.

Janaki (Mythily) Kumar, a User Experience designer at SAP, provided valuable comments on real world practices.

Thomas Kelleher (TJ), my editor at Basic Books for this revised edition, provided rapid, efficient advice and editing suggestions (which led me to yet another massive revision of the manuscript that vastly improved the book). Doug Sery served as my editor at MIT Press for the UK edition of this book (as well as for *Living with Complexity*). For this book, TJ did all the work and Doug provided encouragement.

GENERAL READINGS AND NOTES

In the notes below, I first provide general readings. Then, chapter by chapter, I give the specific sources used or cited in the book.

In this world of rapid access to information, you can find information about the topics discussed here by yourself. Here is an example: In Chapter 5, I discuss root cause analysis as well as the Japanese method called the Five Whys. Although my descriptions of these concepts in Chapter 5 are self-sufficient for most purposes, readers who wish to learn more can use their favorite search engine with the critical phrases in quotes.

Most of the relevant information can be found online. The problem is that the addresses (URLs) are ephemeral. Today's locations of valuable information may no longer be at the same place tomorrow. The creaky, untrustworthy Internet, which is all we have today, may finally, thank goodness, be replaced by a superior scheme. Whatever the reason, the Internet addresses I provide may no longer work. The good news is that over the years that will pass after the publication of this book, new and improved search methods will certainly arise. It should be even easier to find more information about any of the concepts discussed in this book.

These notes provide excellent starting points. I provide critical references for the concepts discussed in the book, organized by the chapters where they were discussed. The citations serve two purposes. First, they provide credit to the originators of the ideas. Second, they serve as starting points to get a deeper understanding of the concepts. For more advanced information (as well as newer, further developments), go out and search. Enhanced search skills are important tools for success in the twenty-first century.

GENERAL READINGS

When the first edition of this book was published, the discipline of interaction design did not exist, the field of human-computer interaction was in its infancy, and most studies were done under the guise of "usability" or "user interface." Several very different disciplines were struggling to bring clarity to this enterprise, but often with little or no interaction among the disciplines. The academic disciplines of computer science, psychology, human factors, and ergonomics all knew of one another's existence and often worked together, but design was not included. Why not design? Note that all the disciplines just listed are in the areas of science and engineering—in other words, technology. Design was then mostly taught in schools of art or architecture as a profession rather than as a research-based academic discipline. Designers had remarkably little contact with science and engineering. This meant that although many excellent practitioners were trained, there was essentially no theory: design was learned through apprenticeship, mentorship, and experience.

Few people in the academic disciplines were aware of the existence of design as a serious enterprise, and as a result, design, and in particular, graphical, communication, and industrial design worked completely independently of the newly emerging discipline of human-computer interaction and the existing disciplines of human factors and ergonomics. Some product design was taught in departments of mechanical engineering, but again, with little interaction with design. Design was simply not an academic discipline, so there was little or no mutual awareness or collaboration. Traces of this distinction remain today, although design is more and more becoming a research-based discipline, where professors have experience in practice as well as PhDs. The boundaries are disappearing.

This peculiar history of many independent, disparate groups all working on similar issues makes it difficult to provide references that cover both the academic side of interaction and experience design, and the applied side of design. The proliferation of books, texts, and journals in human-computer interaction, experience design, and usability is huge: too large to cite. In the materials that follow, I provide a very restricted number of examples. When I originally put together a list of works I considered important, it was far too long. It fell prey to the problem described by Barry Schwartz in his book *The Paradox of Choice: Why More Is Less* (2005). So I decided to simplify by providing less. It is easy to find other works, including important ones that will be published after this book. Meanwhile, my apologies to my many friends whose important and useful works had to be trimmed from my list.

Industrial designer Bill Moggridge was extremely influential in establishing interaction within the design community. He played a major role in the design of the first portable computer. He was one of the three founders of IDEO, one of the world's most influential design firms. He wrote two books of interviews with key people in the early development of the discipline: *Designing Interactions* (2007) and *Designing Media* (2010). As is typical of discussions from the discipline of design, his works focus almost entirely upon the practice of design, with little attention to the science. Barry Katz, a design professor at San Francisco's California College of the Arts, Stanford's d.school, and an IDEO Fellow, provides an excellent history of design practice within the community of companies in Silicon Valley, California: *Ecosystem of Innovation: The History of Silicon Valley Design* (2014). An excellent, extremely comprehensive history of the field of product design is provided by Bernhard Bürdek's *Design: History, Theory, and Practice of Product Design* (2005). Bürdek's book, originally published in German but with an excellent English translation, is the most comprehensive history of product design I have been able to find. I highly recommend it to those who want to understand the historical foundations.

Modern designers like to characterize their work as providing deep insight into the fundamentals of problems, going far beyond the popular conception of design as making things pretty. Designers emphasize this aspect of their profession by discussing the special way in which they approach problems, a method they have characterized as "design thinking." A good introduction to this comes from the book *Change by Design* (2009), by Tim Brown and Barry Katz. Brown is CEO of IDEO and Katz an IDEO Fellow (see the previous paragraph).

An excellent introduction to design research is provided in Jan Chipchase and Simon Steinhardt's *Hidden in Plain Sight* (2013). The book chronicles the life of a design researcher who studies people by observing them in their homes, barber shops, and living quarters around the world. Chipchase is executive creative director of global insights at Frog Design, working out of the Shanghai office. The work of Hugh Beyer and Karen Holtzblatt in *Contextual Design: Defining Customer-Centered Systems* (1998) presents a powerful method of analyzing behavior; they have also produced a useful workbook (Holtzblatt, Wendell, & Wood, 2004).

There are many excellent books. Here are a few more:

Buxton, W. (2007). *Sketching user experience: Getting the design right and the right design*. San Francisco, CA: Morgan Kaufmann. (And see the companion workbook [Greenberg, Carpendale, Marquardt, & Buxton, 2012].)

Coates, D. (2003). *Watches tell more than time: Product design, information, and the quest for elegance*. New York: McGraw-Hill.

Cooper, A., Reimann, R., & Cronin, D. (2007). *About face 3: The essentials of interaction design*. Indianapolis, IN: Wiley Pub.

Hassenzahl, M. (2010). *Experience design: Technology for all the right reasons*. San Rafael, California: Morgan & Claypool.

Moggridge, B. (2007). *Designing interactions*. Cambridge, MA: MIT Press. http://www.designinginteractions.com. Chapter 10 describes the methods of interaction design: http://www.designinginteractions.com/chapters/10

Two handbooks provide comprehensive, detailed treatments of the topics in this book:

Jacko, J. A. (2012). *The human-computer interaction handbook: Fundamentals, evolving technologies, and emerging applications* (3rd edition). Boca Raton, FL: CRC Press.

Lee, J. D., & Kirlik, A. (2013). *The Oxford handbook of cognitive engineering.* New York: Oxford University Press.

Which book should you look at? Both are excellent, and although expensive, well worth the price for anyone who intends to work in these fields. The *Human-Computer Interaction Handbook,* as the title suggests, focuses primarily on computer-enhanced interactions with technology, whereas the *Handbook of Cognitive Engineering* has a much broader coverage. Which book is better? That depends upon what problem you are working on. For my work, both are essential.

Finally, let me recommend two websites:

Interaction Design Foundation: Take special note of its Encyclopedia articles. www.interaction-design.org

SIGCHI: The Computer-Human Interaction Special Interest Group for ACM. www.sigchi.org

CHAPTER ONE: THE PSYCHOPATHOLOGY OF EVERYDAY THINGS

2 *Coffeepot for Masochists:* This was created by the French artist Jacques Carelman (1984). The photograph shows a coffeepot inspired by Carelman, but owned by me. Photograph by Aymin Shamma for the author.

10 *Affordances:* The perceptual psychologist J. J. Gibson invented the word *affordance* to explain how people navigated the world (Gibson, 1979). I introduced the term into the world of interaction design in the first edition of this book (Norman, 1988). Since then, the number of writings on affordance has been enormous. Confusion over the appropriate way to use the term prompted me to introduce the concept of "signifier" in my book *Living with Complexity* (Norman, 2010), discussed throughout this book, but especially in Chapters 1 and 4.

CHAPTER TWO: THE PSYCHOLOGY OF EVERYDAY ACTIONS

38 *Gulfs of execution and evaluation:* The story of the gulfs and bridges of execution and evaluation came from research performed with Ed Hutchins and Jim Hollan, then part of a joint research team between the Naval Personnel Research and Development Center and the University of California, San Diego (Hollan and Hutchins are now professors of

cognitive science at the University of California, San Diego). The work examined the development of computer systems that were easier to learn and easier to use, and in particular, of what has been called direct manipulation computer systems. The initial work is described in the chapter "Direct Manipulation Interfaces" in the book from our laboratories, *User Centered System Design: New Perspectives on Human-Computer Interaction* (Hutchins, Hollan, & Norman, 1986). Also see the paper by Hollan, Hutchins, and David Kirsh, "Distributed Cognition: A New Foundation for Human-Computer Interaction Research" (Hollan, Hutchins, & Kirsh, 2000).

43 *Levitt:* "People don't want to buy a quarter-inch drill. They want a quarter-inch hole!" See Christensen, Cook, & Hal, 2006. The fact that Harvard Business School marketing professor Theodore Levitt is credited with the quote about the drill and the hole is a good example of Stigler's law: "No scientific discovery is named after its original discoverer." Thus, Levitt himself attributed the statement about drills and holes to Leo McGinneva (Levitt, 1983). Stigler's law is, itself, an example of the law: Stigler, a professor of statistics, wrote that he learned the law from the sociologist Robert Merton. See more at Wikipedia, "Stigler's Law of Eponymy" (Wikipedia contributors, 2013c).

46 *Doorknob:* The question "In the house you lived in three houses ago, as you entered the front door, was the doorknob on the left or right?" comes from my paper "Memory, Knowledge, and the Answering of Questions" (Norman, 1973).

53 *Visceral, behavioral, and reflective:* Daniel Kahneman's book, *Thinking Fast and Slow* (Kahneman, 2011), gives an excellent introduction to modern conceptions of the role of conscious and subconscious processing. The distinctions between visceral, behavioral, and reflective processing form the basis of my book *Emotional Design* (Norman, 2002, 2004). This model of the human cognitive and emotional system is described in more technical detail in the scientific paper I wrote with Andrew Ortony and William Revelle: "The Role of Affect and Proto-affect in Effective Functioning" (Ortony, Norman, & Revelle, 2005). Also see "Designers and Users: Two Perspectives on Emotion and Design" (Norman & Ortony, 2006). *Emotional Design* contains numerous examples of the role of design at all three levels.

58 *Thermostat:* The valve theory of the thermostat is taken from Kempton, a study published in the journal *Cognitive Science* (1986). Intelligent thermostats try to predict when they will be required, turning on or off earlier than the simple control illustrated in Chapter 2 can specify, to ensure that the desired temperature is reached at the desired time, without over- or undershooting the target.

63 *Positive psychology:* Mihaly Csikszentmihalyi's work on flow can be found in his several books on the topic (1990, 1997). Martin (Marty) Seligman developed the concept of learned helplessness, and then applied it to depression (Seligman, 1992). However, he decided that it was wrong for

psychology to continually focus upon difficulties and abnormalities, so he teamed up with Csikszentmihalyi to create a movement for positive psychology. An excellent introduction is provided in the article by the two of them in the journal *American Psychologist* (Seligman & Csikszentmihalyi, 2000). Since then, positive psychology has expanded to include books, journals, and conferences.

66 *Human error:* People blame themselves: Unfortunately, blaming the user is imbedded in the legal system. When major accidents occur, official courts of inquiry are set up to assess the blame. More and more often, the blame is attributed to "human error." But in my experience, human error usually is a result of poor design: why was the system ever designed so that a single act by a single person could cause calamity? An important book on this topic is Charles Perrow's *Normal Accidents* (1999). Chapter 5 of this book provides a detailed examination of human error.

72 *Feedforward:* Feedforward is an old concept from control theory, but I first encountered it applied to the seven stages of action in the paper by Jo Vermeulen, Kris Luyten, Elise van den Hoven, and Karin Coninx (2013).

CHAPTER THREE: KNOWLEDGE IN THE HEAD AND IN THE WORLD

74 *American coins:* Ray Nickerson and Marilyn Adams, as well as David Rubin and Theda Kontis, showed that people could neither recall nor recognize accurately the pictures and words on American coins (Nickerson & Adams, 1979; Rubin & Kontis, 1983).

80 *French coins:* The quotation about the French government release of the 10-franc coin comes from an article by Stanley Meisler (1986), reprinted with permission of the *Los Angeles Times*.

80 *Descriptions in memory:* The suggestion that memory storage and retrieval is mediated through partial descriptions was put forth in a paper with Danny Bobrow (Norman & Bobrow, 1979). We argued that, in general, the required specificity of a description depends on the set of items among which a person is trying to distinguish. Memory retrieval can therefore involve a prolonged series of attempts during which the initial retrieval descriptions yield incomplete or erroneous results, so that the person must keep trying, each retrieval attempt coming closer to the answer and helping to make the description more precise.

83 *Constraints of rhyming:* Given just the cues for meaning (the first task), the people David C. Rubin and Wanda T. Wallace tested could guess the three target words used in these examples only 0 percent, 4 percent, and 0 percent of the time, respectively. Similarly, when the same target words were cued only by rhymes, they still did quite poorly, guessing the targets correctly only 0 percent, 0 percent, and 4 percent of the time, respectively. Thus, each cue alone offered little assistance. Combining the meaning cue with the rhyming cue led to perfect performance: the people got the target words 100 percent of the time (Rubin & Wallace, 1989).

86 *'Ali Baba:* Alfred Bates Lord's work is summarized in his book *The Singer of Tales* (1960). The quotation from "'Ali Baba and the Forty Thieves" comes from *The Arabian Nights: Tales of Wonder and Magnificence,* selected and edited by Padraic Colum, translated by Edward William Lane (Colum & Ward, 1953). The names here are in an unfamiliar form: most of us know the magic phrase as "Open Sesame," but according to Colum, "Simsim" is the authentic transliteration.

87 *Passwords:* How do people cope with passwords? There are lots of studies: (Anderson, 2008; Florêncio, Herley, & Coskun, 2007; National Research Council Steering Committee on the Usability, Security, and Privacy of Computer Systems, 2010; Norman, 2009; Schneier, 2000).

To find the most common passwords, just search using some phrase such as "most common passwords." My article on security, which led to numerous newspaper column references to it, is available on my website and was also published in the magazine for human-computer interaction, *Interactions* (Norman, 2009).

89 *Hiding places:* The quotation about professional thieves' knowledge of how people hide things comes from Winograd and Soloway's study "On Forgetting the Locations of Things Stored in Special Places" (1986).

93 *Mnemonics:* Mnemonic methods were covered in my book *Memory and Attention,* and although that book is old, the mnemonic techniques are even older, and are still unchanged (Norman, 1969, 1976). I discuss the effort of retrieval in *Learning and Memory* (Norman, 1982). Mnemonic techniques are easy to find: just search the web for "mnemonics." Similarly, the properties of short- and long-term memory are readily found by an Internet search or in any text on experimental psychology, cognitive psychology, or neuropsychology (as opposed to clinical psychology) or a text on cognitive science. Alternatively, search online for "human memory," "working memory," "short-term memory" or "long-term memory." Also see the book by Harvard psychologist Daniel Schacter, *The Seven Sins of Memory* (2001). What are Schacter's seven sins? Transience, absent-mindedness, blocking, misattribution, suggestibility, persistence, and bias.

101 *Whitehead:* Alfred North Whitehead's quotation about the power of automated behavior is from Chapter 5 of his book *An Introduction to Mathematics* (1911).

107 *Prospective memory:* Considerable research on prospective memory and memory for the future is summarized in the articles by Dismukes on prospective memory and the review by Cristina Atance and Daniela O'Neill on memory for the future, or what they call "episodic future thinking" (Atance & O'Neill, 2001; Dismukes, 2012).

112 *Transactive memory:* The term *transactive memory* was coined by Harvard professor of psychology Daniel Wegner (Lewis & Herndon, 2011; Wegner, D. M., 1987; Wegner, T. G., & Wegner, D. M., 1995).

113 *Stove controls:* The difficulty in mapping stove controls to burners has been understood by human factors experts for over fifty years: Why are

stoves still designed so badly? This issue was addressed in 1959, the very first year of the *Human Factors Journal* (Chapanis & Lindenbaum, 1959).

118 *Culture and design:* My discussion of the impact of culture on mappings was heavily informed by my discussions with Lera Boroditsky, then at Stanford University, but now in the cognitive science department at the University of California, San Diego. See her book chapter "How Languages Construct Time" (2011). Studies of the Australian Aborigine were reported by Núñez & Sweetser (2006).

CHAPTER FOUR: KNOWING WHAT TO DO: CONSTRAINTS, DISCOVERABILITY, AND FEEDBACK

126 *InstaLoad:* A description of Microsoft's InstaLoad technology for battery contacts is available on its website: www.microsoft.com/hardware/en-us/support/licensing-instaload-overview.

129 *Cultural frames:* See Roger Schank and Robert B. Abelson's *Scripts, Plans, Goals, and Understanding* (1977) or Erving Goffman's classic and extremely influential books *The Presentation of Self in Everyday Life* (1959) and *Frame Analysis* (1974). I recommend *Presentation* as the most relevant (and easiest to read) of his works.

129 *Violating social conventions:* "Try violating cultural norms and see how uncomfortable that makes you and the other people." Jan Chipchase and Simon Steinhardt's *Hidden in Plain Sight* provides many examples of how design researchers can deliberately violate social conventions so as to understand how a culture works. Chipchase reports an experiment in which able-bodied young people request that seated subway passengers give up their seat to them. The experimenters were surprised by two things. First, a large proportion of people obeyed. Second, the people most affected were the experimenters themselves: they had to force themselves to make the requests and then felt bad about it for a long time afterward. A deliberate violation of social constraints can be uncomfortable for both the violator and the violated (Chipchase & Steinhardt, 2013).

137 *Light switch panel:* For the construction of my home light switch panel, I relied heavily on the electrical and mechanical ingenuity of Dave Wargo, who actually did the design, construction, and installation of the switches.

156 *Natural sounds:* Bill Gaver, now a prominent design researcher at Goldsmiths College, University of London (UK), first alerted me to the importance of natural sounds in his PhD dissertation and later publications (Gaver, W., 1997; Gaver, W. W., 1989). There has been considerable research on sound since the early days: see, for example, Gygi & Shafiro (2010).

160 *Electric vehicles:* The quotation from the US government rule on sounds for electric vehicles can be found on the Department of Transportation's website (2013).

CHAPTER FIVE: HUMAN ERROR? NO, BAD DESIGN

There has been a lot of work on the study of error, human reliability, and resilience. A good source, besides the items cited below, is the Wiki of Science article on human error (Wiki of Science, 2013). Also see the book *Behind Human Error* (Woods, Decker, Cook, Johannesen, & Sarter, 2010).

Two of the most important workers in human error are British psychologist James Reason and Danish engineer Jens Rasmussen. Also see the books by the Swedish investigator Sidney Dekker, and MIT professor Nancy Leveson (Dekker, 2011, 2012, 2013; Leveson, N., 2012; Leveson, N. G., 1995; Rasmussen, Duncan, & Leplat, 1987; Rasmussen, Pejtersen, & Goodstein, 1994; Reason, J. T., 1990, 2008).

Unless otherwise noted, all the examples of slips in this chapter were collected by me, primarily from the errors of myself, my research associates, my colleagues, and my students. Everyone diligently recorded his or her slips, with the requirement that only the ones that had been immediately recorded would be added to the collection. Many were first published in Norman (1981).

165 *F-22 crash:* The analysis of the Air Force F-22 crash comes from a government report (Inspector General United States Department of Defense, 2013). (This report also contains the original Air Force report as Appendix C.)

170 *Slips and mistakes:* The descriptions of skill-based, rule-based, and knowledge-based behavior is taken from Jens Rasmussen's paper on the topic (1983), which still stands as one of the best introductions. The classification of errors into slips and mistakes was done jointly by me and Reason. The classification of mistakes into rule-based and knowledge-based follows the work of Rasmussen (Rasmussen, Goodstein, Andersen, & Olsen, 1988; Rasmussen, Pejtersen, & Goodstein, 1994; Reason, J. T., 1990, 1997, 2008). Memory lapse errors (both slips and mistakes) were not originally distinguished from other errors: they were put into separate categories later, but not quite the same way I have done here.

172 *"Gimli Glider":* The so-called Gimli Glider accident was an Air Canada Boeing 767 that ran out of fuel and had to glide to a landing at Gimli, a decommissioned Canadian Air Force base. There were numerous mistakes: search for "Gimli Glider accident." (I recommend the Wikipedia treatment.)

174 *Capture error:* The category "capture error" was invented by James Reason (1979).

178 *Airbus:* The difficulties with the Airbus and its modes are described in (Aviation Safety Network, 1992; Wikipedia contributors, 2013a). For a disturbing description of another design problem with the Airbus—

that the two pilots (the captain and the first officer) can both control the joysticks, but there is no feedback, so one pilot does not know what the other pilot is doing—see the article in the British newspaper *The Telegraph* (Ross & Tweedie, 2012).

181 *The Kiss nightclub fire in Santa Maria, Brazil:* It is described in numerous Brazilian and American newspapers (search the web for "Kiss nightclub fire"). I first learned about it from the *New York Times* (Romero, 2013).

186 *Tenerife crash:* My source for information about the Tenerife crash is from a report by Roitsch, Babcock, and Edmunds issued by the American Airline Pilots Association (Roitsch, Babcock, & Edmunds, undated). It is perhaps not too surprising that it differs in interpretation from the Spanish government's report (Spanish Ministry of Transport and Communications, 1978), which in turn differs from the report by the Dutch Aircraft Accident Inquiry Board. A nice review of the 1977 Tenerife accident—written in 2007—that shows its long-lasting importance has been written by Patrick Smith for the website Salon.com (Smith, 2007, Friday, April 6, 04:00 AM PDT).

188 *Air Florida crash:* The information and quotations about the Air Florida crash are from the report of the National Transportation Safety Board (1982). See also the two books entitled *Pilot Error* (Hurst, 1976; Hurst, R. & Hurst, L. R., 1982). The two books are quite different. The second is better than the first, in part because at the time the first book was written, not much scientific evidence was available.

190 *Checklists in medicine:* Duke University's examples of knowledge-based mistakes can be found at Duke University Medical Center (2013). An excellent summary of the use of checklists in medicine—and the many social pressures that have slowed up its adoption—is provided by Atul Gawande (2009).

192 *Jidoka:* The quotation from Toyota about *Jidoka,* and the Toyota Production System comes from the auto maker's website (Toyota Motor Europe Corporate Site, 2013). Poka-yoke is described in many books and websites. I found the two books written by or with the assistance of the originator, Shigeo Shingo, to provide a valuable perspective (Nikkan Kogyo Shimbun, 1988; Shingo, 1986).

193 *Aviation safety:* The website for NASA's Aviation Safety Reporting System provides details of the system, along with a history of its reports (NASA, 2013).

197 *Hindsight:* Baruch Fischhoff's study is called "Hindsight ≠ Foresight: The Effect of Outcome Knowledge on Judgment Under Uncertainty" (1975). And while you are at it, see his more recent work (Fischhoff, 2012; Fischhoff & Kadvany, 2011).

198 *Designing for error:* I discuss the idea of designing for error in a paper in *Communications of the ACM,* in which I analyze a number of the slips people make in using computer systems and suggest system design principles that might minimize those errors (Norman, 1983). This philosophy also pervades the book that our research team put together:

User Centered System Design (Norman & Draper, 1986); two chapters are especially relevant to the discussions here: my "Cognitive Engineering" and the one I wrote with Clayton Lewis, "Designing for Error."

200 *Multitasking:* There are many studies of the dangers and inefficiencies of multitasking. A partial review is given by Spink, Cole, & Waller (2008). David L. Strayer and his colleagues at the University of Utah have done numerous studies demonstrating rather severe impairment in driving behavior while using cell phones (Strayer & Drews, 2007; Strayer, Drews, & Crouch, 2006). Even pedestrians are distracted by cell phone usage, as demonstrated by a team of researchers from West Washington University (Hyman, Boss, Wise, McKenzie, & Caggiano, 2010).

200 *Unicycling clown:* The clever study of the invisible clown, riding a unicycle, "Did you see the unicycling clown? Inattentional blindness while walking and talking on a cell phone" was done by Hyman, Boss, Wise, McKenzie, & Caggiano (2010).

208 *Swiss cheese model:* James Reason introduced his extremely influential Swiss cheese model in 1990 (Reason, J., 1990; Reason, J. T., 1997).

210 *Hersman:* Deborah Hersman's description of the design philosophy for aircraft comes from her talk on February 7, 2013, discussing the NTSB's attempts to understand the cause of the fires in the battery compartments of Boeing 787 aircraft. Although the fires caused airplanes to make emergency landings, no passengers or crew were injured: the multiple layers of redundant protection maintained safety. Nonetheless, the fires and resulting damage were unexpected and serious enough that all Boeing 787 airlines were grounded until all parties involved had completed a thorough investigation of the causes of the incident and then gone through a new certification process with the Federal Aviation Agency (for the United States, and through the corresponding agencies in other countries). Although this was expensive and greatly inconvenient, it is an example of good proactive practice: take measures before accidents lead to injury and death (National Transportation Safety Board, 2013).

212 *Resilience engineering:* The excerpt from "Prologue: Resilience Engineering Concepts," in the book *Resilience Engineering,* is reprinted by permission of the publishers (Hollnagel, Woods, & Leveson, 2006).

213 *Automation:* Much of my research and writings have addressed issues of automation. An early paper, "Coffee Cups in the Cockpit," addresses this problem as well as the fact that when talking about incidents in a large country—or that occur worldwide—a "one-in-a-million chance" is not good enough odds (Norman, 1992). My book *The Design of Future Things* deals extensively with this issue (Norman, 2007).

214 Royal Majesty *accident:* An excellent analysis of the mode error accident with the cruise ship *Royal Majesty* is contained in Asaf Degani's book on automation, *Taming HAL: Designing Interfaces Beyond 2001* (Degani, 2004), as well as in the analyses by Lützhöft and Dekker and the official NTSB report (Lützhöft & Dekker, 2002; National Transportation Safety Board, 1997).

CHAPTER SIX: DESIGN THINKING

As pointed out in the "General Readings" section, a good introduction to design thinking is *Change by Design* by Tim Brown and Barry Katz (2009). Brown is CEO of IDEO and Katz a professor at the California College of the Arts, visiting professor at Stanford's d.school, and an IDEO Fellow. There are multiple Internet sources; I like designthinkingforeducators.com.

220 *Double diverge-converge pattern:* The double diverge-converge pattern was first introduced by the British Design Council in 2005, which called it the "Double-Diamond Design Process Model" (Design Council, 2005).

221 *HCD process:* The human-centered design process has many variants, each similar in spirit but different in the details. A nice summary of the method I describe is provided by the HCD book and toolkit from the design firm IDEO (IDEO, 2013).

227 *Prototyping:* For prototyping, see Buxton's book and handbook on sketching (Buxton, 2007; Greenberg, Carpendale, Marquardt, & Buxton, 2012). There are multiple methods used by designers to understand the nature of the problem and come to a potential solution. Vijay Kumar's *101 Design Methods* (2013) doesn't even cover them all. Kumar's book is an excellent treatment of design research methods, but its focus is on innovation, not the production of products, so it does not cover the actual development cycle. Physical prototyping, their tests, and iterations are outside the domain, as are the practical concerns of the marketplace, the topic of the last part of this chapter and all of chapter 7.

227 *Wizard of Oz technique:* The Wizard of Oz technique is named after L. Frank Baum's book *The Wonderful Wizard of Oz* (Baum & Denslow, 1900). My use of the technique is described in the resulting paper from the group headed by artificial intelligence researcher Danny Bobrow at what was then called the Xerox Palo Alto Research Center (Bobrow et al., 1977). The "graduate student" sitting in the other room was Allen Munro, who then went on to a distinguished research career.

229 *Nielsen:* Jakob Nielsen's argument that five users is the ideal number for most tests can be found on the Nielsen Norman group's website (Nielsen, 2013).

233 *Three goals:* Marc Hassenzahl's use of the three levels of goals (be-goals, do-goals, and motor-goals) is described in many places, but I strongly recommend his book *Experience Design* (Hassenzahl, 2010). The three goals come from the work of Charles Carver and Michael Scheier in their landmark book on the use of feedback models, chaos, and dynamical theory to explain much of human behavior (Carver & Scheier, 1998).

246 *Age and performance:* A good review of the impact of age on human factors is provided by Frank Schieber (2003). The report by Igo Grossman and

colleagues is a typical example of research showing that careful studies reveal superior performance with age (Grossmann et al., 2010).

254 *Swatch International Time:* Swatch's development of .beat time and the French decimal time are discussed in the Wikipedia article on decimal time (Wikipedia contributors, 2013b).

CHAPTER SEVEN: DESIGN IN THE WORLD OF BUSINESS

261 *Creeping featurism:* A note for the technology historians. I've managed to trace the origin of this term to a talk by John Mashey in 1976 (Mashey, 1976). At that time Mashey was a computer scientist at Bell Laboratories, where he was one of the early developers of UNIX, a well-known computer operating system (which is still active as Unix, Linux, and the kernel underlying Apple's Mac OS).

262 *Youngme Moon:* Youngme Moon's book *Different: Escaping the Competitive Herd* (Moon, 2010) argues that "If there is one strain of conventional wisdom pervading every company in every industry, it is the importance of competing hard to differentiate yourself from the competition. And yet going head-to-head with the competition—with respect to features, product augmentations, and so on—has the perverse effect of making you just like everyone else." (From the jacket of her book: see http://youngmemoon.com/Jacket.html.)

266 *Word-gesture system:* The word-gesture system that works by tracing the letters on the screen keyboard to type rapidly and efficiently (although not as fast as with a traditional ten-finger keyboard) is described in considerable detail by Shumin Zhai and Per Ola Kristensson, two of the developers of this method of typing (Zhai & Kristensson, 2012).

269 *Multitouch screens:* In the more than thirty years multitouch screens have been in the laboratories, numerous companies have launched products and failed. Nimish Mehta is credited with the invention of multitouch, discussed in his master's thesis (1982) from the University of Toronto. Bill Buxton (2012), one of the pioneers in this field, provides a valuable review (he was working with multitouch displays in the early 1980s at the University of Toronto). Another excellent review of multitouch and gestural systems in general (as well as design principles) is provided by Dan Saffer in his book *Designing Gestural Interfaces* (2009). The story of Fingerworks and Apple is readily found by searching the web for "Fingerworks."

270 *Stigler's law:* See the comment about this in the notes for Chapter 2.

271 *Telephonoscope:* The illustration of the "Telephonoscope" was originally published in the December 9, 1878, issue of the British magazine *Punch* (for its 1879 Almanack). The picture comes from Wikipedia (Wikipedia contributors, 2013d), where it is in the public domain because of its age.

276 *QWERTY keyboard:* The history of the QWERTY keyboard is discussed in numerous articles. I thank Professor Neil Kay of University of Strathclyde for our e-mail correspondence and his article "Rerun the

Tape of History and QWERTY Always Wins" (2013). This article led me to the "QWERTY People Archive" website by the Japanese researchers Koichi and Motoko Yasuoka, an incredibly detailed, valuable resource for those interested in the history of the keyboard, and in particular, of the QWERTY configuration (Yasuoka & Yasuoka, 2013). The article on the typewriter in the 1872 *Scientific American* is fun to read: the style of *Scientific American* has changed drastically since then (Anonymous, 1872).

278 *Dvorak keyboard:* Is Dvorak faster than QWERTY? Yes, but not by much: Diane Fisher and I studied a variety of keyboard layouts. We thought that alphabetically organized keys would be superior for beginners. No, they weren't: we discovered that knowledge of the alphabet was not useful in finding the keys. Our studies of alphabetical and Dvorak keyboards were published in the journal *Human Factors* (Norman & Fisher, 1984).

Admirers of the Dvorak keyboard claim much more than a 10 percent improvement, as well as faster learning rates and less fatigue. But I will stick by my studies and my statements. If you want to read more, including a worthwhile treatment of the history of the typewriter, see the book *Cognitive Aspects of Skilled Typewriting*, edited by William E. Cooper, which includes several chapters of research from my laboratory (Cooper, W. E., 1963; Norman & Fisher, 1984; Norman & Rumelhart, 1963; Rumelhart & Norman, 1982).

278 *Keyboard ergonomics:* Health aspects of keyboards are reported in National Institute of Health (2013).

279 *Incremental and radical innovation:* The Italian business professor Roberto Verganti and I discuss the principles of incremental and radical innovation (Norman & Verganti, 2014; Verganti, 2009, 2010).

281 *Hill climbing:* There are very good descriptions of the hill-climbing process for design in Christopher Alexander's book *Notes on the Synthesis of Form* (1964) and Chris Jones's book *Design Methods* (1992; also see Jones, 1984).

286 *Humans versus machines:* The remarks by MIT professor Erik Brynjolfsson were made in his talk at the June 2012 National Academy of Engineering symposium on manufacturing, design, and innovation (Brynjolfsson, 2012). His book, coauthored with Andrew McAfee—*Race Against the Machine: How the Digital Revolution Is Accelerating Innovation, Driving Productivity, and Irreversibly Transforming Employment and the Economy*—contains an excellent treatment of design and innovation (Brynjolfsson & McAfee, 2011).

290 *Interactive media:* Al Gore's interactive media book is *Our Choice* (2011). Some of the videos from my early interactive book are still available: see Norman (1994 and 2011b).

295 *Rise of the small:* The section "The Rise of the Small" is taken from my essay written for the hundredth anniversary of the Steelcase company, reprinted here with Steelcase's permission (Norman, 2011a).

REFERENCES

Alexander, C. (1964). *Notes on the synthesis of form*. Cambridge, England: Harvard University Press.

Anderson, R. J. (2008). *Security engineering—A guide to building dependable distributed systems* (2nd edition). New York, NY: Wiley. http://www.cl.cam.ac.uk/~rja14/book.html

Anonymous. (1872). The type writer. *Scientific American, 27*(6, August 10), 1.

Atance, C. M., & O'Neill, D. K. (2001). Episodic future thinking. *Trends in Cognitive Sciences, 5*(12), 533–537. http://www.sciencessociales.uottawa.ca/ccll/eng/documents/15Episodicfuturethinking_000.pdf

Aviation Safety Network. (1992). Accident description: Airbus A320-111. Retrieved February 13, 2013, from http://aviation-safety.net/database/record.php?id=19920120-0

Baum, L. F., & Denslow, W. W. (1900). *The wonderful wizard of Oz*. Chicago, IL; New York, NY: G. M. Hill Co. http://hdl.loc.gov/loc.rbc/gen.32405

Beyer, H., & Holtzblatt, K. (1998). *Contextual design: Defining customer-centered systems*. San Francisco, CA: Morgan Kaufmann.

Bobrow, D., Kaplan, R., Kay, M., Norman, D., Thompson, H., & Winograd, T. (1977). GUS, a frame-driven dialog system. *Artificial Intelligence, 8*(2), 155–173.

Boroditsky, L. (2011). How Languages Construct Time. In S. Dehaene & E. Brannon (Eds.), *Space, time and number in the brain: Searching for the foundations of mathematical thought*. Amsterdam, The Netherlands; New York, NY: Elsevier.

Brown, T., & Katz, B. (2009). *Change by design: How design thinking transforms organizations and inspires innovation*. New York, NY: Harper Business.

Brynjolfsson, E. (2012). Remarks at the June 2012 National Academy of Engineering symposium on Manufacturing, Design, and Innovation. In

K. S. Whitefoot & S. Olson (Eds.), *Making value: Integrating manufacturing, design, and innovation to thrive in the changing global economy*. Washington, DC: The National Academies Press.

Brynjolfsson, E., & McAfee, A. (2011). *Race against the machine: How the digital revolution is accelerating innovation, driving productivity, and irreversibly transforming employment and the economy*. Lexington, MA: Digital Frontier Press (Kindle Edition). http://raceagainstthemachine.com/

Bürdek, B. E. (2005). *Design: History, theory, and practice of product design*. Boston, MA: Birkhäuser–Publishers for Architecture.

Buxton, W. (2007). *Sketching user experience: Getting the design right and the right design*. San Francisco, CA: Morgan Kaufmann.

Buxton, W. (2012). Multi-touch systems that I have known and loved. Retrieved February 13, 2013, from http://www.billbuxton.com/multi-touchOverview.html

Carelman, J. (1984). *Catalogue d'objets introuvables: Et cependant indispensables aux personnes telles que acrobates, ajusteurs, amateurs d'art*. Paris, France: Éditions Balland.

Carver, C. S., & Scheier, M. (1998). *On the self-regulation of behavior*. Cambridge, UK; New York, NY: Cambridge University Press.

Chapanis, A., & Lindenbaum, L. E. (1959). A reaction time study of four control-display linkages. *Human Factors, 1*(4), 1–7.

Chipchase, J., & Steinhardt, S. (2013). *Hidden in plain sight: How to create extraordinary products for tomorrow's customers*. New York, NY: HarperCollins.

Christensen, C. M., Cook, S., & Hal, T. (2006). What customers want from your products. *Harvard Business School Newsletter: Working Knowledge*. Retrieved February 2, 2013, from http://hbswk.hbs.edu/item/5170.html

Coates, D. (2003). *Watches tell more than time: Product design, information, and the quest for elegance*. New York, NY: McGraw-Hill.

Colum, P., & Ward, L. (1953). *The Arabian nights: Tales of wonder and magnificence*. New York, NY: Macmillan. (Also see http://www.bartleby.com/16/905.html for a similar rendition of *'Ali Baba and the Forty Thieves*.)

Cooper, A., Reimann, R., & Cronin, D. (2007). *About face 3: The essentials of interaction design*. Indianapolis, IN: Wiley.

Cooper, W. E. (Ed.). (1963). *Cognitive aspects of skilled typewriting*. New York, NY: Springer-Verlag.

Csikszentmihalyi, M. (1990). *Flow: The psychology of optimal experience*. New York, NY: Harper & Row.

Csikszentmihalyi, M. (1997). *Finding flow: The psychology of engagement with everyday life*. New York, NY: Basic Books.

Degani, A. (2004). Chapter 8: The grounding of the *Royal Majesty*. In A. Degani (Ed.), *Taming HAL: Designing interfaces beyond 2001*. New York, NY: Palgrave Macmillan. http://ti.arc.nasa.gov/m/profile/adegani/Grounding%20of%20the%20Royal%20Majesty.pdf

Dekker, S. (2011). *Patient safety:A human factors approach*. Boca Raton, FL: CRC Press.

Dekker, S. (2012). *Just culture: Balancing safety and accountability*. Farnham, Surrey, England; Burlington, VT: Ashgate.

Dekker, S. (2013). *Second victim: Error, guilt, trauma, and resilience*. Boca Raton, FL: Taylor & Francis.

Department of Transportation, National Highway Traffic Safety Administration. (2013). Federal motor vehicle safety standards: Minimum sound requirements for hybrid and electric vehicles. Retrieved from https://www.federalregister.gov/articles/2013/01/14/2013-00359/federal-motor-vehicle-safety-standards-minimum-sound-requirements-for-hybrid-and-electric-vehicles-p-79

Design Council. (2005). The "double-diamond" design process model. Retrieved February 9, 2013, from http://www.designcouncil.org.uk/designprocess

Dismukes, R. K. (2012). Prospective memory in workplace and everyday situations. *Current Directions in Psychological Science 21*(4), 215–220.

Duke University Medical Center. (2013). Types of errors. Retrieved February 13, 2013, from http://patientsafetyed.duhs.duke.edu/module_e/types_errors.html

Fischhoff, B. (1975). Hindsight ≠ foresight: The effect of outcome knowledge on judgment under uncertainty. *Journal of Experimental Psychology: Human Perception and Performance, 104*, 288–299. http://www.garfield.library.upenn.edu/classics1992/A1992HX83500001.pdf is a nice reflection on this paper by Baruch Fischhoff, in 1992. (The paper was declared a "citation classic.")

Fischhoff, B. (2012). *Judgment and decision making*. Abingdon, England; New York, NY: Earthscan.

Fischhoff, B., & Kadvany, J. D. (2011). *Risk: A very short introduction*. Oxford, England; New York, NY: Oxford University Press.

Florêncio, D., Herley, C., & Coskun, B. (2007). Do strong web passwords accomplish anything? Paper presented at Proceedings of the 2nd USENIX workshop on hot topics in security, Boston, MA. http://www.usenix.org/event/hotsec07/tech/full_papers/florencio/florencio.pdf and also http://research.microsoft.com/pubs/74162/hotsec07.pdf

Gaver, W. (1997). Auditory Interfaces. In M. Helander, T. K. Landauer, & P. V. Prabhu (Eds.), *Handbook of human-computer interaction* (2nd, completely rev. ed., pp. 1003–1041). Amsterdam, The Netherlands; New York, NY: Elsevier.

Gaver, W. W. (1989). The SonicFinder: An interface that uses auditory icons. *Human-Computer Interaction, 4*(1), 67–94. http://www.informaworld.com/10.1207/s15327051hci0401_3

Gawande, A. (2009). *The checklist manifesto: How to get things right*. New York, NY: Metropolitan Books, Henry Holt and Company.

Gibson, J. J. (1979). *The ecological approach to visual perception*. Boston, MA: Houghton Mifflin.

Goffman, E. (1959). *The presentation of self in everyday life*. Garden City, NY: Doubleday.

Goffman, E. (1974). *Frame analysis: An essay on the organization of experience*. New York, NY: Harper & Row.

Gore, A. (2011). *Our choice: A plan to solve the climate crisis* (ebook edition). Emmaus, PA: Push Pop Press, Rodale, and Melcher Media. http://pushpoppress.com/ourchoice/

Greenberg, S., Carpendale, S., Marquardt, N., & Buxton, B. (2012). *Sketching user experiences: The workbook*. Waltham, MA: Morgan Kaufmann.

Grossmann, I., Na, J., Varnum, M. E. W., Park, D. C., Kitayama, S., & Nisbett, R. E. (2010). Reasoning about social conflicts improves into old age. *Proceedings of the National Academy of Sciences.* http://www.pnas.org/content/early/2010/03/23/1001715107.abstract

Gygi, B., & Shafiro, V. (2010). *From signal to substance and back: Insights from environmental sound research to auditory display design* (Vol. 5954). Berlin & Heidelberg, Germany: Springer. http://link.springer.com/chapter/10.1007%2F978-3-642-12439-6_16?LI=true

Hassenzahl, M. (2010). *Experience design: Technology for all the right reasons.* San Rafael, CA: Morgan & Claypool.

Hollan, J. D., Hutchins, E., & Kirsh, D. (2000). Distributed cognition: A new foundation for human-computer interaction research. *ACM Transactions on Human-Computer Interaction: Special Issue on Human-Computer Interaction in the New Millennium, 7*(2), 174–196. http://hci.ucsd.edu/lab/hci_papers/JH1999-2.pdf

Hollnagel, E., Woods, D. D., & Leveson, N. (Eds.). (2006). *Resilience engineering: Concepts and precepts.* Aldershot, England; Burlington, VT: Ashgate. http://www.loc.gov/catdir/toc/ecip0518/2005024896.html

Holtzblatt, K., Wendell, J., & Wood, S. (2004). *Rapid contextual design: A how-to guide to key techniques for user-centered design*. San Francisco, CA: Morgan Kaufmann.

Hurst, R. (1976). *Pilot error: A professional study of contributory factors.* London, England: Crosby Lockwood Staples.

Hurst, R., & Hurst, L. R. (1982). *Pilot error: The human factors* (2nd edition). London, England; New York, NY: Granada.

Hutchins, E., J., Hollan, J., & Norman, D. A. (1986). Direct manipulation interfaces. In D. A. Norman & S. W. Draper (Eds.), *User centered system design; New perspectives on human-computer interaction* (pp. 339–352). Mahwah, NJ: Lawrence Erlbaum Associates.

Hyman, I. E., Boss, S. M., Wise, B. M., McKenzie, K. E., & Caggiano, J. M. (2010). Did you see the unicycling clown? Inattentional blindness while walking and talking on a cell phone. *Applied Cognitive Psychology, 24*(5), 597–607. http://dx.doi.org/10.1002/acp.1638

IDEO. (2013). Human-centered design toolkit. IDEO website. Retrieved February 9, 2013, from http://www.ideo.com/work/human-centered-design-toolkit/

Inspector General United States Department of Defense. (2013). *Assessment of the USAF aircraft accident investigation board (AIB) report on the F-22A mishap of November 16, 2010.* Alexandria, VA: The Department of Defense Office of the Deputy Inspector General for Policy and Oversight. http://www.dodig.mil/pubs/documents/DODIG-2013-041.pdf

Jacko, J. A. (2012). *The human-computer interaction handbook: Fundamentals, evolving technologies, and emerging applications* (3rd edition.). Boca Raton, FL: CRC Press.

Jones, J. C. (1984). *Essays in design*. Chichester, England; New York, NY: Wiley.

Jones, J. C. (1992). *Design methods* (2nd edition). New York, NY: Van Nostrand Reinhold.

Kahneman, D. (2011). *Thinking, fast and slow*. New York, NY: Farrar, Straus and Giroux.

Katz, B. (2014). *Ecosystem of innovation: The history of Silicon Valley design*. Cambridge, MA: MIT Press.

Kay, N. (2013). Rerun the tape of history and QWERTY always wins. *Research Policy*.

Kempton, W. (1986). Two theories of home heat control. *Cognitive Science, 10,* 75–90.

Kumar, V. (2013). *101 design methods: A structured approach for driving innovation in your organization*. Hoboken, NJ: Wiley. http://www.101designmethods.com/

Lee, J. D., & Kirlik, A. (2013). *The Oxford handbook of cognitive engineering*. New York: Oxford University Press.

Leveson, N. (2012). *Engineering a safer world*. Cambridge, MA: MIT Press. http://mitpress.mit.edu/books/engineering-safer-world

Leveson, N. G. (1995). *Safeware: System safety and computers*. Reading, MA: Addison-Wesley.

Levitt, T. (1983). *The marketing imagination*. New York, NY; London, England: Free Press; Collier Macmillan.

Lewis, K., & Herndon, B. (2011). Transactive memory systems: Current issues and future research directions. *Organization Science, 22*(5), 1254–1265.

Lord, A. B. (1960). *The singer of tales*. Cambridge, MA: Harvard University Press.

Lützhöft, M. H., & Dekker, S. W. A. (2002). On your watch: Automation on the bridge. *Journal of Navigation, 55*(1), 83–96.

Mashey, J. R. (1976). Using a command language as a high-level programming language. Paper presented at *Proceedings of the 2nd international conference on Software engineering*, San Francisco, California, USA.

Mehta, N. (1982). *A flexible machine interface*. M.S. Thesis, Department of Electrical Engineering, University of Toronto.

Meisler, S. (1986, December 31). Short-lived coin is a dealer's delight. *Los Angeles Times*, 1–7.

Moggridge, B. (2007). *Designing interactions*. Cambridge, MA: MIT Press. http://www.designinginteractions.com—Chapter 10 describes the methods of interaction design: http://www.designinginteractions.com/chapters/10

Moggridge, B. (2010). *Designing media*. Cambridge, MA: MIT Press.

Moon, Y. (2010). *Different: Escaping the competitive herd*. New York, NY: Crown Publishers.

NASA, A. S. R. S. (2013). NASA Aviation Safety Reporting System. Retrieved February 19, 2013, from http://asrs.arc.nasa.gov

National Institute of Health. (2013). PubMed Health: Carpal tunnel syndrome. From http://www.ncbi.nlm.nih.gov/pubmedhealth/PMH0001469/

National Research Council Steering Committee on the Usability Security and Privacy of Computer Systems. (2010). *Toward better usability, security, and privacy of information technology: Report of a workshop.* The National Academies Press. http://www.nap.edu/openbook.php?record_id=12998

National Transportation Safety Board. (1982). *Aircraft accident report: Air Florida, Inc., Boeing 737-222, N62AF, collision with 14th Street Bridge near Washington National Airport (Executive Summary).* NTSB Report No. AAR-82-08. http://www.ntsb.gov/investigations/summary/AAR8208.html

National Transportation Safety Board. (1997). *Marine accident report grounding of the Panamanian passenger ship ROYAL MAJESTY on Rose and Crown Shoal near Nantucket, Massachusetts June 10, 1995* (NTSB Report No. MAR-97-01, adopted on 4/2/1997): National Transportation Safety Board. Washington, DC. http://www.ntsb.gov/doclib/reports/1997/mar9701.pdf

National Transportation Safety Board. (2013). NTSB Press Release: NTSB identifies origin of JAL Boeing 787 battery fire; design, certification and manufacturing processes come under scrutiny. Retrieved February 16, 2013, from http://www.ntsb.gov/news/2013/130207.html

Nickerson, R. S., & Adams, M. J. (1979). Long-term memory for a common object. *Cognitive Psychology, 11*(3), 287–307. http://www.sciencedirect.com/science/article/pii/0010028579900136

Nielsen, J. (2013). Why you only need to test with 5 users. Nielsen Norman group website. Retrieved February 9, 2013, from http://www.nngroup.com/articles/why-you-only-need-to-test-with-5-users/

Nikkan Kogyo Shimbun, Ltd. (Ed.). (1988). *Poka-yoke: Improving product quality by preventing defects.* Cambridge, MA: Productivity Press.

Norman, D. A. (1969, 1976). *Memory and attention: An introduction to human information processing* (1st, 2nd editions). New York, NY: Wiley.

Norman, D. A. (1973). Memory, knowledge, and the answering of questions. In R. Solso (Ed.), *Contemporary issues in cognitive psychology: The Loyola symposium.* Washington, DC: Winston.

Norman, D. A. (1981). Categorization of action slips. *Psychological Review, 88*(1), 1–15.

Norman, D. A. (1982). *Learning and memory.* New York, NY: Freeman.

Norman, D. A. (1983). Design rules based on analyses of human error. *Communications of the ACM, 26*(4), 254–258.

Norman, D. A. (1988). *The psychology of everyday things.* New York, NY: Basic Books. (Reissued in 1990 [Garden City, NY: Doubleday] and in 2002 [New York, NY: Basic Books] as *The design of everyday things.*)

Norman, D. A. (1992). Coffee cups in the cockpit. In *Turn signals are the facial expressions of automobiles* (pp. 154–174). Cambridge, MA: Perseus Publishing. http://www.jnd.org/dn.mss/chapter_16_coffee_c.html

Norman, D. A. (1993). *Things that make us smart.* Cambridge, MA: Perseus Publishing.

Norman, D. A. (1994). *Defending human attributes in the age of the machine*. New York, NY: Voyager. http://vimeo.com/18687931
Norman, D. A. (2002). Emotion and design: Attractive things work better. *Interactions Magazine, 9*(4), 36–42. http://www.jnd.org/dn.mss/Emotion-and-design.html
Norman, D. A. (2004). *Emotional design: Why we love (or hate) everyday things*. New York, NY: Basic Books.
Norman, D. A. (2007). *The design of future things*. New York, NY: Basic Books.
Norman, D. A. (2009). When security gets in the way. *Interactions, 16*(6), 60–63. http://jnd.org/dn.mss/when_security_gets_in_the_way.html
Norman, D. A. (2010). *Living with complexity*. Cambridge, MA: MIT Press.
Norman, D. A. (2011a). The rise of the small. *Essays in honor of the 100th anniversary of Steelcase*. From http://100.steelcase.com/mind/don-norman/
Norman, D. A. (2011b). Video: Conceptual models. Retrieved July 19, 2012, from http://www.interaction-design.org/tv/conceptual_models.html
Norman, D. A., & Bobrow, D. G. (1979). Descriptions: An intermediate stage in memory retrieval. *Cognitive Psychology, 11*, 107–123.
Norman, D. A., & Draper, S. W. (1986). *User centered system design: New perspectives on human-computer interaction*. Mahwah, NJ: Lawrence Erlbaum Associates.
Norman, D. A., & Fisher, D. (1984). Why alphabetic keyboards are not easy to use: Keyboard layout doesn't much matter. *Human Factors, 24*, 509–519.
Norman, D. A., & Ortony, A. (2006). Designers and users: Two perspectives on emotion and design. In S. Bagnara & G. Crampton-Smith (Eds.), *Theories and practice in interaction design* (pp. 91–103). Mahwah, NJ: Lawrence Erlbaum Associates.
Norman, D. A., & Rumelhart, D. E. (1963). Studies of typing from the LNR Research Group. In W. E. Cooper (Ed.), *Cognitive aspects of skilled typewriting*. New York, NY: Springer-Verlag.
Norman, D. A., & Verganti, R. (in press, 2014). Incremental and radical innovation: Design research versus technology and meaning change. *Design Issues*. http://www.jnd.org/dn.mss/incremental_and_radi.html
Núñez, R., & Sweetser, E. (2006). With the future behind them: Convergent evidence from Aymara language and gesture in the crosslinguistic comparison of spatial construals of time. *Cognitive Science, 30*(3), 401–450.
Ortony, A., Norman, D. A., & Revelle, W. (2005). The role of affect and proto-affect in effective functioning. In J.-M. Fellous & M. A. Arbib (Eds.), *Who needs emotions? The brain meets the robot* (pp. 173–202). New York, NY: Oxford University Press.
Oudiette, D., Antony, J. W., Creery, J. D., & Paller, K. A. (2013). The role of memory reactivation during wakefulness and sleep in determining which memories endure. *Journal of Neuroscience, 33*(15), 6672.
Perrow, C. (1999). *Normal accidents: Living with high-risk technologies*. Princeton, NJ: Princeton University Press.
Portigal, S., & Norvaisas, J. (2011). Elevator pitch. *Interactions, 18*(4, July), 14–16. http://interactions.acm.org/archive/view/july-august-2011/elevator-pitch1

Rasmussen, J. (1983). Skills, rules, and knowledge: Signals, signs, and symbols, and other distinctions in human performance models. *IEEE Transactions on Systems, Man, and Cybernetics, SMC-13*, 257–266.

Rasmussen, J., Duncan, K., & Leplat, J. (1987). *New technology and human error*. Chichester, England; New York, NY: Wiley.

Rasmussen, J., Goodstein, L. P., Andersen, H. B., & Olsen, S. E. (1988). *Tasks, errors, and mental models: A festschrift to celebrate the 60th birthday of Professor Jens Rasmussen*. London, England; New York, NY: Taylor & Francis.

Rasmussen, J., Pejtersen, A. M., & Goodstein, L. P. (1994). *Cognitive systems engineering*. New York, NY: Wiley.

Reason, J. T. (1979). Actions not as planned. In G. Underwood & R. Stevens (Eds.), *Aspects of consciousness*. London: Academic Press.

Reason, J. (1990). The contribution of latent human failures to the breakdown of complex systems. *Philosophical Transactions of the Royal Society of London. Series B, Biological Sciences 327*(1241), 475–484.

Reason, J. T. (1990). *Human error*. Cambridge, England; New York, NY: Cambridge University Press.

Reason, J. T. (1997). *Managing the risks of organizational accidents*. Aldershot, England; Brookfield, VT: Ashgate.

Reason, J. T. (2008). *The human contribution: Unsafe acts, accidents and heroic recoveries*. Farnham, England; Burlington, VT: Ashgate.

Roitsch, P. A., Babcock, G. L., & Edmunds, W. W. (undated). *Human factors report on the Tenerife accident*. Washington, DC: Air Line Pilots Association. http://www.skybrary.aero/bookshelf/books/35.pdf

Romero, S. (2013, January 27). Frenzied scene as toll tops 200 in Brazil blaze. *New York Times*, from http://www.nytimes.com/2013/01/28/world/americas/brazil-nightclub-fire.html?_r=0 Also see: http://thelede.blogs.nytimes.com/2013/01/27/fire-at-a-nightclub-in-southern-brazil/?ref=americas

Ross, N., & Tweedie, N. (2012, April 28). Air France Flight 447: "Damn it, we're going to crash." *The Telegraph*, from http://www.telegraph.co.uk/technology/9231855/Air-France-Flight-447-Damn-it-were-going-to-crash.html

Rubin, D. C., & Kontis, T. C. (1983). A schema for common cents. *Memory & Cognition, 11*(4), 335–341. http://dx.doi.org/10.3758/BF03202446

Rubin, D. C., & Wallace, W. T. (1989). Rhyme and reason: Analyses of dual retrieval cues. *Journal of Experimental Psychology: Learning, Memory, and Cognition, 15*(4), 698–709.

Rumelhart, D. E., & Norman, D. A. (1982). Simulating a skilled typist: A study of skilled cognitive-motor performance. *Cognitive Science, 6*, 1–36.

Saffer, D. (2009). *Designing gestural interfaces*. Cambridge, MA: O'Reilly.

Schacter, D. L. (2001). *The seven sins of memory: How the mind forgets and remembers*. Boston, MA: Houghton Mifflin.

Schank, R. C., & Abelson, R. P. (1977). *Scripts, plans, goals, and understanding: An inquiry into human knowledge structures*. Hillsdale, NJ: L. Erlbaum Associates; distributed by the Halsted Press Division of John Wiley and Sons.

Schieber, F. (2003). Human factors and aging: Identifying and compensating for age-related deficits in sensory and cognitive function. In N. Charness & K. W. Schaie (Eds.), *Impact of technology on successful aging* (pp. 42–84). New York, NY: Springer Publishing Company. http://sunburst.usd.edu/~schieber/psyc423/pdf/human-factors.pdf

Schneier, B. (2000). *Secrets and lies: Digital security in a networked world.* New York, NY: Wiley.

Schwartz, B. (2005). *The paradox of choice: Why more is less.* New York, NY: HarperCollins.

Seligman, M. E. P. (1992). *Helplessness: On depression, development, and death.* New York, NY: W. H. Freeman.

Seligman, M. E. P., & Csikszentmihalyi, M. (2000). Positive psychology: An introduction. *American Psychologist, 55*(1), 5–14.

Sharp, H., Rogers, Y., & Preece, J. (2007). *Interaction design: Beyond human-computer interaction* (2nd edition). Hoboken, NJ: Wiley.

Shingo, S. (1986). *Zero quality control: Source inspection and the poka-yoke system.* Stamford, CT: Productivity Press.

Smith, P. (2007). Ask the pilot: A look back at the catastrophic chain of events that caused history's deadliest plane crash 30 years ago. Retrieved from http://www.salon.com/2007/04/06/askthepilot227/ on February 7, 2013.

Spanish Ministry of Transport and Communications. (1978). *Report of a collision between PAA B-747 and KLM B-747 at Tenerife, March 27, 1977.* Translation published in *Aviation Week and Space Technology,* November 20 and 27, 1987.

Spink, A., Cole, C., & Waller, M. (2008). Multitasking behavior. *Annual Review of Information Science and Technology, 42*(1), 93–118.

Strayer, D. L., & Drews, F. A. (2007). Cell-phone–induced driver distraction. *Current Directions in Psychological Science, 16*(3), 128–131.

Strayer, D. L., Drews, F. A., & Crouch, D. J. (2006). A Comparison of the cell phone driver and the drunk driver. *Human Factors: The Journal of the Human Factors and Ergonomics Society, 48*(2), 381–391.

Toyota Motor Europe Corporate Site. (2013). Toyota production system. Retrieved February 19, 2013, from http://www.toyota.eu/about/Pages/toyota_production_system.aspx

Verganti, R. (2009). *Design-driven innovation: Changing the rules of competition by radically innovating what things mean.* Boston, MA: Harvard Business Press. http://www.designdriveninnovation.com/

Verganti, R. (2010). User-centered innovation is not sustainable. *Harvard Business Review Blogs* (March 19, 2010). http://blogs.hbr.org/cs/2010/03/user-centered_innovation_is_no.html

Vermeulen, J., Luyten, K., Hoven, E. V. D., & Coninx, K. (2013). Crossing the bridge over Norman's gulf of execution: Revealing feedforward's true identity. Paper presented at CHI 2013, Paris, France.

Wegner, D. M. (1987). Transactive memory: A contemporary analysis of the group mind. In B. Mullen & G. R. Goethals (Eds.), *Theories of group behavior* (pp. 185–208). New York, NY: Springer-Verlag. http://www.wjh.harvard.edu/~wegner/pdfs/Wegner Transactive Memory.pdf

Wegner, T. G., & Wegner, D. M. (1995). Transactive memory. In A. S. R. Manstead & M. Hewstone (Eds.), *The Blackwell encyclopedia of social psychology* (pp. 654–656). Oxford, England; Cambridge, MA: Blackwell.

Whitehead, A. N. (1911). *An introduction to mathematics.* New York, NY: Henry Holt and Company

Wiki of Science (2013). Error (human error). Retrieved from http://wikiofscience.wikidot.com/quasiscience:error on February 6, 2013.

Wikipedia contributors. (2013a). Air Inter Flight 148. *Wikipedia, The Free Encyclopedia.* Retrieved February 13, 2103, from http://en.wikipedia.org/w/index.php?title=Air_Inter_Flight_148&oldid=534971641

Wikipedia contributors. (2013b). Decimal time. *Wikipedia, The Free Encyclopedia.* Retrieved February 13, 2013, from http://en.wikipedia.org/w/index.php?title=Decimal_time&oldid=501199184

Wikipedia contributors. (2013c). Stigler's law of eponymy. *Wikipedia, The Free Encyclopedia.* Retrieved February 2, 2013, from http://en.wikipedia.org/w/index.php?title=Stigler%27s_law_of_eponymy&oldid=531524843

Wikipedia contributors. (2013d). Telephonoscope. *Wikipedia, The Free Encyclopedia.* Retrieved February 8, 2013, from http://en.wikipedia.org/w/index.php?title=Telephonoscope&oldid=535002147

Winograd, E., & Soloway, R. M. (1986). On forgetting the locations of things stored in special places. *Journal of Experimental Psychology: General, 115*(4), 366–372.

Woods, D. D., Dekker, S., Cook, R., Johannesen, L., & Sarter, N. (2010). *Behind human error* (2nd edition). Farnham, Surry, UK; Burlington, VT: Ashgate.

Yasuoka, K., & Yasuoka, M. (2013). QWERTY people archive. Retrieved February 8, 2013, from http://kanji.zinbun.kyoto-u.ac.jp/db-machine/~yasuoka/QWERTY/

Zhai, S., & Kristensson, P. O. (2012). The word-gesture keyboard: Reimagining keyboard interaction. *Communications of the ACM, 55*(9), 91–101. http://www.shuminzhai.com/shapewriter-pubs.htm

INDEX

Abelson, Bob, 129
A/B testing, 224–225
Accidents
  "Five Whys," 165–169
  investigation of, 163–169, 197–198
  root cause analysis of, 164
  social and institutional pressures and, 186–191
  when human error really is to blame, 210–211
  See also Error; Mistakes; Slips
Acoustical memory, 94
Action
  Gulfs of Execution and Evaluation and, 38–40
  opportunistic, 43
  reversing, 199, 203, 205
  stages of, 40–44, 55–56, 71–73, 172–173
  subconscious nature of many, 42
  See also Psychology of everyday actions
Action slips, 171, 173, 174, 194
Activity
  complete immersion into, 55–56
  task *vs.*, 232–234
Activity-centered controls, 140–141
Activity-centered design, 231–234
Adams, Marilyn, 74
Affordances, xiv–xv, 10–13, 19–20, 60, 72, 145, 298
  applying to everyday objects, 132–141
  minimizing chance of inappropriate actions using, 67
  misuse of term, 13–14
  perceived, 13, 18, 19, 145
  signifiers *vs.*, xiv–xv, 14, 18, 19
Agile process of product development, 234
Airbus accident, 178–179
Air Florida crash, 188–189
Airplane
  attitude indicator design, 121–122
  failure of automation in, 214
  landing gear switch design, 135
  mode-error slips and control design, 178–179
  See also Aviation

Airplane accidents, 164–166, 172, 178–179, 186–187, 188–189, 314
Air-traffic control instructions, pilots remembering, 105–107
Alarm clocks, mode-error slips and, 178
Alert, sound signifier as, 160
"'Ali Baba and the Forty Thieves," 86
Altair 8800 computer, 274
Amazon.com, 264
Andon, 192
Annoyance of sounds, 156, 160
Anti-lock brakes, rule-based mistake in using, 182
Apple, 121, 233, 250, 270, 272, 289
Apple QuickTake camera, 272
Arithmetic, mental, 103–104
Automation, 185, 213–214, 248–316
Automobiles
  activity-centered design of, 231–232
  application of constraints to, 202
  auditory and haptic modalities for warning systems, 95
  door handles, 133–134
  failure of first American, 274, 279–280
  incremental innovation in, 279–280
  interlocks and, 142
  limited life span of, 292
  seat adjustment control, 22
  standardization and, 248
  starting, 141–142
  technology and changes in, 267–268
  See also Driving
Aviation
  deliberate violation example, 211
  interruptions and errors in, 200
  NASA's safety reporting system, 193–194
  use of checklists in, 189–190, 191
  See also Airplane
Baby locks, 144
Battery design, 125–127
Baum, L. Frank, 227
Beeps, 156
Be-goals, 233
Behavior
  constraints forcing desired, 141–145
  data-driven, 43
  event-driven, 42, 43
  goal-driven, 42–43, 44
  knowledge-based, 179, 180
  knowledge in the head and in the world and, 75–85
  rule-based, 179, 180
  skill-based, 179, 180, 206–207
  technology accommodating, 68–71
Behavioral level of processing, 51–53
  design and, 54, 55
  emotional responses and, 56
  relation to visceral and reflective stages, 54–55
  stages of action and, 55–56
Bell, Alexander Graham, 270
Benz, Karl, 279
Benz & Cie, 279
Bezos, Jeff, 264
Big data, 224–225
Biometric devices, 128
Blame, for error, 162, 163
  falsely blaming self, 61, 65–71, 167
  misplaced, 61–62
Boats, control of, 21–22
Bookmarks, 16
Books, *see* e-books
Brainstorming, 226
British Design Council, 220
British Psychological Society, 150
Brynjolfsson, Erik, 286–287
Budgets, product development, 237, 240
Business strategy, lock-ins as, 143–144

Cabinet doors, lack of signifiers on, 134
Calendar program, using variety of formats, 70–71
Cameras
  digital, 272, 274
  merger with cell phones, 265
Cane, design of, 245
Capture slips, 174, 208
Carelman, Jacques, 2
Carpal tunnel syndrome, 278
Carver, Charles, 233
Catalogue d'objets introuvables (Carelman), 2
Causal elements, reflective level of processing and, 53
Causes of events
  causal relations, 59–65
  need to form explanations and, 57–59
Cell phones, 34, 200, 265, 280. *See also* Telephone
Celsius scale, conversion between Fahrenheit scale and, 101–102
Change, technology as cause of, 264–268, 282, 284–285
Checklists, 189–191
Chess-playing machine, 286–287
Child safety caps, 144
Chord keyboards, 279
Cisco, 273
Clocks, 249, 250
Clothing industry, yearly changes in fashion, 292
"Coffeepot for Masochists," 2
Cognition and emotion, 49–55
  conscious, 48, 49, 51–52, 53, 100–101
  distributed, 287–288
  integration of, 47, 48–55
  behavioral level, 50, 51–55
  design and levels of, 53–55
  reflective level, 50, 53–55
  stages of action and levels of processing, 55–56
  subconscious, 44–49, 51–52, 173, 206–207
  technology and enhanced human, 285–288
  visceral level, 50–51, 53–55
Coins
  confusion created by new design of, 79–82
  types of knowledge and use of, 74–75, 77, 79–80
Communication
  conceptual models and, 31–32
  design and, 8–9, 73
  technological change and, 283
Companies, conservatism of large, 269
Competition-driven design, 259–264
Complexity, 4–8
  complicated *vs.*, 247
  using conceptual model to tame, 247–248
Conceptual models, 10, 25–37, 40, 72, 94, 96, 98, 121, 204, 298
  communication and, 31–32
  as story, 57–59
  and Gulfs of Evaluation and Execution, 39, 40
  mental models, 26, 31
  providing meaning via, 99–100
  to tame complexity, 247–248
  for thermostat, 57–59, 68–69
Confirmation messages, 203–205
Conscious cognition, 48, 49, 51–52, 53, 100–101
  knowledge-based behavior and, 184
  mistakes and, 173
  subconscious *vs.*, 40, 42, 44–56, 67, 310
Constraints, 10, 73
  applied to everyday objects, 132–141
  to bridge Gulf of Execution, 40
  cultural (*see* Cultural constraints)
  on design process, 240–247

Constraints *(Continued)*
desired behavior and, 76, 141–145
knowledge in the world and, 123, 124–125
logical, 124–125, 130
memory and, 82–85
minimizing chance of inappropriate actions using, 67, 202–203
physical (*see* Physical constraints)
semantic, 124–125, 129–130
signifiers and, 132–135
Consumer economy, 291–293
Controls
activity-centered, 140–141
device-centered, 140
incorporating safety or security in, 256
mapping and design of, 21
segregating, 203
See also Switches
Conventions, cultural. *See* Cultural conventions
Cooperative problem-solving, 185
Cost
as design constraint, 6, 219, 230, 240, 241, 242, 245, 260, 294
feedback design and, 23–25, 68
Countersteering, 102–103
Creativity, 49, 64
Creeping featurism, 258, 261–264
Csikszentmihalyi, Mihaly, 55–56
Cultural constraints, 124–125, 128–129
on assembly of mechanical device, 85
behavior and, 76
cultural conventions and, 130–132, 146
standardization as, 248
Cultural conventions
behavior and, 76
as cultural constraints, 130–132, 146
destination-control elevators and change in, 146–149
faucet design and, 151–152
mapping and, 151–152
people's responses to changes in, 149–150
perceived affordance and, 145
Cultural norms
confusion and lack of knowledge of, 134–135
conventions and standards, 130–132
Culture
impact of technology on, 285
mappings and, 22–23, 118–122
pace of change of, 282
Customers
observing would-be, 222–223, 225–226
quality and focus on, 264
See also Purchasers; Users
Cybermind, 112
Cyborgs, 284

Daily Mail (newspaper), 88
Daimler, 279
Data-driven behavior, 43
Data networks, 281–282
Dead man's switch, 142–143
Decision gates, 234, 235
Declarative knowledge, 78
Declarative memory, 47, 97
Deliberate violations, 211
accidents and, 169–170
Dependence on technology, 285–287
Description, discrimination among choices and, 80–82
Description-similarity slips, 174, 175
Design
activity-centered, 231–234
areas of specialty in, 4–5, 9,110, 302, 308
behavioral level and, 54, 55
challenge of, 34–36, 239–247

checklist, 191
choice of metaphor and, 120–122
coins, of, 79–82
communication and, 8–9, 73
competition-driven, 259–264
constraints as tools for, 85
correct requirements/specifications and, 229–230, 234–235
double-diamond diverge-converge model, 219, 220–221
as equalizing tool, 297
error and (*see* Error)
experience, 4–5, 9, 302, 307
faucet, 115–116, 150–155
flexibility in, 246–247
fundamental principles of, 71–73, 298. *See also individual principles*
implications of short-term memory for, 94–95
inclusive design, 243–247
industrial, 4–5, 9, 302, 306
interaction, 4–5, 9, 306, 309
interplay of technology and psychology in, 6–8
knowledge in the world and the head and, 76–77
legacy problem, 127, 266, 274
management of process, 34–35
memory-lapse mistakes and, 185–186
moral obligations of, 291–293
multidisciplinary approach to, 34–36, 238–239, 242–243
problem identification and, 217–220
providing meaningful structure in, 100
reflection and, 53–54
rule-based mistakes and, 182–183, 184
security and, 90–91, 255–257
success of, 293–294
superfluous features in, 291–293
theory *vs.* practice in, 236–239
universal (inclusive), 243–247
visceral responses and, 51
in the years 1988–2038, 282–288
See also Human-centered design (HCD)
Design error, operator error *vs.*, 6–8
Designers
advice for, 64–65
bridging Gulfs of Evaluation and Execution, 40
clients/customers, 240–241
conceptual model and, 31–32
engineers as, 6–8, 10
The Design of Future Things (Norman), 185
Design redundancy, 210
Design research
market research *vs.*, 224–226
observation, 222–224
separating from product team, 238–239
Design team, 35
multidisciplinary, 34–36, 238–239, 242–243
needs of other groups in product process, 241–242
Design thinking, 219, 293–298
double-diamond diverge-converge model of design, 219, 220–221
See also Human-centered design (HCD)
Destination-control elevators, 146–149
Detection of error, 194–198
Development cycle, 260, 268–279
Device-centered controls, 140
Different (Moon), 262–263
Digital cameras, 272, 274
Digital picture frame, 272
Digital time, 252–254
Digital watch, 27–28, 33
Discoverability, 72, 298
affordances, 10–13, 19–20
conceptual models, 25–31
constraints, 10

Discoverability *(Continued)*
design and, 3–4
feedback, 23–25
gesture-controlled devices and, 115–116
mappings, 20–23
signifiers, 13–20
Discrimination, rules for, 80–82
Displays, 68
description-similarity slips and, 175
mapping and design, 21
metaphor and interaction with, 120–122
smart, 121, 265–266
touch-sensitive, 21, 140, 268–269
Distributed cognition, 287–288
Do-goals, 233
Doors
affordances and, 3,13–16, 18, 69, 132–135, 145
designing for security, 255
handles/hardware, 18, 133–134, 145
panic bars, 60, 133
poor design of, 1–3
signifiers and, 14–16, 18, 132–135
sliding, 16
Double-diamond diverge-converge model of design, 219, 220–221
Drill, goal of buying, 43–44
Driver's safety device, 142–143
Driving
cell phone use while, 200
conventions of, 131–132
left-side *vs.* right-side, 122
as rule-based behavior, 181
stages of action in, 40–41
sterile periods during, 200–201
while drunk, 211
See also Automobiles
du Maurier, George, 270–271
Durable goods, 291
Duryea, 274, 280
Dvorak, August, 278
Dvorak keyboard, 278
Early adopters, 271
Edison, Thomas, 270
Electrical standards, 249
e-Books (Electronic books), 16, 143, 286, 288–290, 319
Electronic games, 282
Electronic reminders, 109
Elevators, destination-control, 146–149
Emotion, xiii, xv, 5, 47–56, 293–295, 310
behavioral level, 50–56
cognition and, 47–50, 53–55
positive and negative, 10, 38, 49, 63–64
reflective level, 50, 53–56
visceral level, 50–51, 53–56
Emotional Design (Norman), 49, 54
Engineers
as designers, 6–8, 10
as users of design team output, 241–242
Environment, attributing failure/error to, 61–62, 63, 168
Environmental cue, as reminder, 109
Epic poems, memory for, 82–85
Error, 66–68, 162–216
automation and, 213–214
checklist to reduce, 189–191
classification as slips or mistakes, 170
defined, 170–171
deliberate violations and, 169–170
design and, 162–163, 198–211, 215–216
design to prevent or lessen cost of, 67–68, 198–210, 202–205
detecting, 194–198
reasons for, 163–169
reporting, 191–194
resilience engineering and, 211–213
social and institutional pressures and, 186–191
See also Mistakes; Slips

Error messages, 203–205
Ethnography, 222–224
Evaluation, 38–40, 216
  action cycle and stages of, 40–44
Event-driven behavior, 42, 43
Everyday practice, scientific theory
    vs., 104–105
Execution, 38–40, 216
  action cycle and stages of, 40–44
  feedforward information and,
    71–72
Expectations
  behavioral cognition and, 52
  emotions and, 52–53
Experience design, 4–5, 9, 302, 307
Experts
  design and, 6
  Jidoka and, 192
  slips and, 7, 173, 199
  unconscious action and, 47,
    100–101, 173, 180, 216
Eyewitness testimony, 97

Fahrenheit scale, conversion
    between Celsius scale and,
    101–102
Failure
  attributing reason for, 61–62
"fail frequently, fail fast," 229
  learned helplessness and, 62–63
  learning from, 64, 229
  positive psychology and, 63–65
  self-blame and, 65–71, 113,
    162–169
Farber, Sam, 244–245
Faucet design, 115–116, 150–155
Featuritis, xvii, 258, 261–265
Federal Aviation Authority (FAA),
    193–194, 200
Federal Communications
    Commission (FCC), 250, 251
Feedback, 10, 23–25, 298
  as aid in design, 71–72
  behavioral states and, 52
  to bridge Gulf of Evaluation,
    39, 40
  characteristics of effective, 23–24
  communicating progress, 60
  faucet design and, 153
  prioritizing, 25
  reducing error and, 216
Feedforward, 71–72, 216
Filing cabinet, Gulfs of Evaluation
    and Execution and, 37–39
Financial institutions, mistake
    outcomes, 198
Financial transactions, sensibility
    checks and, 206
Fingerworks, 269–270
Fire exit lockout, 144
Fire extinguisher pins, 144
Fischhoff, Baruch, 197
"Five Whys" analysis, 165–169, 219
Flexibility, designing to
    accommodate, 246–247
Flow state, 55–56
Forcing functions, 141–142, 143
  deliberate disabling of, 145
  interlocks, 142–143
  lock-ins, 143–144
  lockouts, 144–145
  memory-lapse slips and,
    176–177
  reducing error and, 216
Ford, Henry, 292
Foresight ≠ hindsight, 197, 315
Frames, 129
Freud, Sigmund, 173
F-22 airplane accidents, 164–166

Games, 256
Gated product development
    methods, 234, 235
General Electric, 30
Generalizations, forming, 57
Gestalt psychology, 12, 22
Gestural keyboards, 278
Gesture-controlled faucets, soap
    dispensers and hand dryers,
    115–116
Gibson, J. J., 12
Gibsonian psychology, 12

Gimli Glider Air Canada 767 accident, 172, 314
Global Positioning System (GPS), 214, 281
Goal
be-goal, do-goal, and motor-goal, 233
comparing outcome with, 41
conscious *vs.* unconscious, 42
stages of execution, 41, 42–43
Goal-driven behavior, 42–43, 44
Goffman, Erving, 129
Google, 90
Gore, Al, 290
GPS. *See* Global Positioning System (GPS)
Graphical user interface, 100
Greetings, cultural conventions regarding, 130–131
Gulf of Evaluation, 38–40, 216
Gulf of Execution, 38–40, 216

Hand dryers, gesture-controlled, 115–116
Handed-up technology, 297
Haptics, 95
Hassenzahl, Marc, 233
HCD. *See* Human-centered design (HCD)
Hersman, Deborah, 210
High-definition television (HDTV), 250–252, 272
Highway signs, misinterpreting, 196–197
Hill climbing, 281
Hindsight,
explanations given in, 183, 197–198, 315
foresight ≠ to, 197, 315
Hollnagel, Erik, 212
Homer's *Odyssey* and *Iliad*, 84
Household appliances, 240–241, 292
Human-centered design (HCD), 8–10, 137, 219–220, 221–236
activity-centered design *vs.*, 231–234
design thinking and, 219
idea generation (ideation) in, 222, 226–227
incremental innovation and, 281
iteration, 229–230, 234–236
iterative design *vs.* linear stages, 234–236
observation/design research and, 222–226
in practice, 236–239
prototyping in, 222, 227–228
role of, 9–10
spiral method, 222. *See also* Iteration
testing in, 222, 228–229
Human error, *See* Error
Human-machine interaction, 6, 185, 215
Hutchins, Edwin, 287
HyperCard, 289

Idea generation (ideation), 222, 226–227
Identity theft, 90
IDEO, 64, 229, 303, 307
"fail frequently, fail fast," 229
"if only" statements, accidents and, 209
Iliad (Homer), 84
Implanted devices, 284
Implicit knowledge, 236
Inclusive design, 243–247
Incremental innovation, 279–281
Individual
as focus of design, 231, 233
technology and empowerment of, 295–297
Industrial design, 4–5, 9
Industrial Design Society of America (IDSA), 5
Industrial settings, natural mapping and, 117
Information pickup, 12

Innovation, xvii, 43, 374, 279–282, 397, 317
radical and incremental, 279–282, 319
Inside-out display, 121–122
InstaLoad battery contacts (Microsoft), 126, 127, 313
Institutional pressure, accidents and, 186–191
Instruction manuals, *see* manuals
Interaction, principles of, xii–319
Interlocks, 142–143
Interpret, in action cycle, 41
Interruptions, as source of error, 163, 176, 199–200
iPod, 233
Iteration in design, 222, 229–230, 234–236. *See also* Repetitive cycles

Jidoka, 192
Joysticks, 21
Junghans Mega 1000 digital watch, 27–28

KAIST, wall at, 18
Kasparov, Gary, 287
Kelly, David, 229
Key
automobile, 141–142
physical constraints and design of, 127–128
Keyboard, evolution of, 264–267, 274–279, 318–319. *See also* QWERTY
Key logger, 91
Kiss nightclub fire, 181
Kitchen organization, 247
KLM Boeing 747 crash, 186–187
Knobs, 13, 177
Knowledge
arbitrary, 98–100
declarative, 78
procedural, 78–79
retrieval of, 97–98
Knowledge-based behavior, 179, 180
Knowledge-based mistakes, 171–172, 184–185
Knowledge in the head, 74–75, 105–109, 123
behavior and, 75–77, 79–85
memory as, 86–91
in multiple heads, multiple devices, 111–113
prospective memory and, 107–109
remembering air-traffic control instructions and, 105–107
tradeoff with knowledge in the world, 109–111
Knowledge in the world, 74–75, 77–79, 123
behavior and, 75–79
Lego motorcycle construction and, 123–125
operating technology and, 216
tradeoff with knowledge in the head, 109–111
See also Constraints
Kuhn Rikon, 244

Law, cultural convention codified into, 131
"Law of Product Development," xvii, 237–239, 261
Learned helplessness, 62–63
Learned skills, 51–53
Learning
changes in convention and new, 149–150
conscious thinking and, 45–46, 100–101
failure and, 64
knowledge in the environment and, 78
rote, 98
Legacy problem, 127, 266, 274
Lego motorcycle, 123–125, 129, 130, 262, 263

Leveson, Nancy, 212
Levitt, Theodore, 43–44
Life cycle, product, 294
Light, stages of turning on, 40, 42
Light controls, activity-centered, 140–141
Light, as feedback, 23–24
Light switches, mapping and, 20–21, 135–140
Linear stages of design, 234–236
Living with Complexity (Norman), 14, 247
Lizard brain, 50–51
Location-based reminders, 109
Lock-ins, 143–144
Lockouts, 144–145
Locks, physical constraints and design of, 127–128
Logical constraints, 124–125, 130
Long-term memory (LTM), 47, 95–98
Lord, Albert Bates, 83–84

Machine-people interaction, 68, 185, 215
Machine-readable codes, 207
Machines, characteristics of, 5–6
Management, role in design, 34–35
Management review, 234, 235
Manuals, 3–4, 26, 27, 29, 180, 185, 294
  system image and, 31
Manufacturing, product success and, 294
Mapping, 10, 20–23, 72, 298
  bridging Gulf of Execution and, 40
  culture and, 118–122
  faucet design and, 151, 154
  levels of, 115
  minimizing chance of inappropriate actions using, 67
  natural (*see* Natural mapping)
Market analytics, 224–225
Marketing
  effect on design, 277–278
  product success and, 294
Market research, design research *vs.*, 224–226
McAfee, Andrew, 287
Meaning, semantic constraints and, 129–130
Meaningful things, memory for, 98–100
Medicine
  checklists in, 190–191
  electronic records, 95
  errors in, 198, 200, 206
  interruptions in, 200
  safety reporting system, 194
Memory
  acoustical, 94
  approximate methods and, 100–105
  for arbitrary things, 98–100
  constraints and, 82–85
  declarative, 47, 97
  distortions/falsification in, 96
  knowledge in the head and, 86–91, 105–109
  long-term, 47, 95–98
  for meaningful things, 98–100
  in multiple heads, multiple devices, 111–113
  procedural, 47, 96–97
  prospective, 107–109
  reflective, 53–54
  retrieval, 45–47
  short-term (working), 92–95
  structure of, 91–105
  transactive, 111–112
  use of mnemonics, 88, 93–94, 99
  See also Knowledge in the head
Memory-lapse mistakes, 171, 172, 185–186, 195, 199–200
Memory-lapse slips, 171, 173, 176–177, 195, 199–200
Mental arithmetic, 103–104
Mental models, 26, 31. Conceptual models
Mercedes-Benz, 22, 279
Metaphor, design and choice of, 120–122

Metric measurement, 149, 253, 254
  accidents resulting from conversion, 172, 314
Microsoft
  flexible date and time formats, 70–71
  InstaLoad battery contacts, 126, 127, 313
Microwave ovens, interlocks and, 142
Mistakes, 170–173
  classification of, 179–186
  confirmation messages and, 204–205
  detecting, 194, 195
  explaining away, 195–196
  knowledge-based, 171–172, 184–185
  memory-lapse, 171, 172, 185–186, 195
  rule-based, 171, 180–184
  See also Error; Slips
Mitsubishi, 269
Mnemonics, 88, 93–94, 99
Mode error slips, 174, 177–179, 207
Models
  approximate, 100–105
  See also Conceptual models
Modes, 177–178
Moon, Youngme, 262–263
Moral obligations of design, 291–293
Motorcycle
  Lego, 123–125, 129, 130, 262, 263
  steering system, 102–103
  turn signal switch, 99–100
Motor-goal, 233
Motor system, visceral response and, 50–51
Multidisciplinary approach to design, 34–36, 238–239, 242–243
Multitasking, error and, 200
Multitouch displays, 269, 270
Music, technological change and, 283

Names
  identifying people by, 89–90
  memory for, 98
Narrative, conceptual models as form of, 57–59
National Academy of Engineering, 286
National Aeronautics and Space Administration (NASA), 193–194
National Highway and Traffic Safety Administration (NHTSA), 157, 159–160
National Institute of Health (NIH), 278
National Transportation Safety Board (NTSB), 135, 188–189, 198, 210
Natural mapping, 22, 113–118. *See also* Mapping
  culture and, 118–122
  gesture-controlled devices and, 115–116
  in industrial settings, 117
  as knowledge in the world, 79
  light switches and, 137–140
  reducing error and, 216
  spatial cues and, 115
  stove controls and, 113–115, 116–117, 118
  tradeoffs, 117–118
Negative emotional state, 49
Nest thermostat, 68–69
Nickerson, Ray, 74
Nielsen, Jakob, 229
Nielsen Norman group, 303, 317
Nissan, 158
Nonstandard clock, 249, 250
Norman, Don 92
"Norman doors," 1–3
Norman's law of product development, xvii, 237–239, 261, 309 310

Norms, cultural, 130–132
Novices, mistakes and, 173, 199
NTSB. *See* National Transportation Safety Board (NTSB)
Nuclear power plant accident, 7, 201

Observation, in human-centered design, 222–226
Odyssey (Homer), 84
Office copiers, design constraint for, 241
Our Choice (Gore), 290
Outside-in display, 121, 122
Overlearning, 45–46
OXO, 244–245

Paller, Ken, 96
Palo Alto Research Center (PARC), 227, 317
Panic bars, 60, 133
Paris Métro doors, 134–135
Passwords, remembering, 86–89, 91, 312
Patents, 238
Pedestrians, and electric cars, 157–161
Penny, knowledge in the head and in the world and, 74–75, 77
People with special needs, designing for, 243–247
Perceive, as stage of evaluation, 41
Perceived affordances, 13, 18, 19, 145. *See also* Signifiers
Perform, as stage of execution, 41
Personality, attributing failure to, 61–62
Physical anthropometry, 243
Physical constraints, 124–128
  battery design and, 125–127
  forcing functions, 141–142, 143
  as knowledge in the world, 79
  locks and keys and, 127–128
Pilots, remembering air-traffic control instructions, 105–107
Plan, as stage of execution, 41
Planned obsolescence, 291–292
Plato, 286
Poetry, constraints of, 82–85
Poka-yoke, 193
Porsche, 158
Positive psychology, 63–65
Precision, knowledge and, 76, 79–82
Predictive typing, 266
Price, design and competition/ focus on, 241, 259, 260, 264
Problem identification in design, 217–220
  double-diamond diverge-converge model of design and, 220–221
  See also Human-centered design (HCD)
Problem solving, reflective, 46–47
Procedural knowledge, 78–79
Procedural memory, 47, 96–97
Product development
  competitive forces in, 259–264
  cycle of, 268–279
  Don Norman's law of, 237–239
  managing, 235–236
  multidisciplinary needs, 34–36, 238–239, 241–243
  process of, 221–230, 234–236
  prototyping, 227–228
  technology and, 258, 264–268
  timing of innovation, 271–272
Product manager, 230
Products
  development cycle, 260, 268–279
  failure of new products, 272, 274
  life cycle of, 294
  stage gate methods, 234, 235
  success of, 293–294
Prospective memory, 107–109
Prototyping, 222, 227–228, 235
Psychology, 27–73. *See also* Cognition
  causal relations (blame), 59–65
  cognition and emotion, 49–55. *See also* Cognition; Emotion

falsely blaming self, 65–71
fundamental design principles and, 71–73
Gibsonian, 12
Gulfs of Evaluation and Execution, 38–40
human thought, 44–49
interplay with technology, 6–8
people as storytellers, 56–59
positive, 63–65
stages of action, 40–44, 55–56, 71–73, 172–173
The Psychology of Everyday Things (POET), xi, 283, 299–304
Punch (magazine), 270
Purchasers
designing for, 241
users *vs.*, 117–118
See also Customers
Purchasing process, usability and, 117–118

Quality, focus on customer and, 264
Questioning, 46, 117, 226–227, 229, 230, 264, 286, 295, 310
QWERTY keyboard, 254, 266, 275–278, 318, 319. *See also* Keyboard

Radiation doses, sensibility checks and, 206
Radical innovation, 279–280, 281–282
Rasmussen, Jens, 179
Reading *vs.* listening, 267
Reason, James, 164, 170, 208
Recycling, 294
Reflection, 45
design and, 53–54
relation to visceral and behavioral response, 54–55
stages of action and, 55–56
Reflective problem solving, 46–47
Refrigerator temperature controls, conceptual model and, 28–31
Rehearsal of material, 96, 100–101
Reminders, 108–109
Reminding, strategies for, 106, 107–109, 110
Remington typewriter, 275, 276, 277
Remote controller, cultural effect on design of, 118, 119
Repetitive cycles of design, *see* spiral design
Resilience engineering, 211–213
Retention, memory and, 94
Retrieval, memory and, 97–98
Retrospective decision making, 183
Reversing action, *see* Undo
Rhyming, constraints of, 83
Root cause analysis, 42, 43–44, 164–165
Rote learning, 98
Royal Majesty cruise ship, 214
Rubin, David, 83
Rule-based behavior, 179, 180
Rule-based mistakes, 171, 180–184
Rules, deliberate violation of, 169–170

Safety. *See also* Accidents; Error
checklists, 189–191
electric vehicles, and 157–161
forcing functions, 142–145
interlocks, 142–143
lock-ins, 143–144
lockouts and, 144–145
management, 209–210, 212–213
NASA's safety reporting system, 193–194
resilience engineering, 211–213
social and institutional pressures and, 186–189
sterile periods and, 200
Swiss cheese metaphor, 208–210
warning signals, 201
Sayeki, Yutaka, 99–100, 102–103, 105
Schank, Roger, 128–129
Schedules, product development, 237, 240

Scheier, Michael, 233
Schindler elevators, 147
Scripts, 129
Scrum method of product development, 234
Security
  design and, 90–91, 255–257
  identity theft and, 90
  passwords as means of ensuring, 86–89, 91
Semantic constraints, 124–125, 129–130
Sensibility checks, 199, 205–206
Shingo, Shigeo, 193
Sholes, Christopher Latham
  typewriter, 275–276. *See also* QWERTY
Short-term memory (STM), 92–95, 102
Shower controls, design of, 73
Signifiers, xv, 10, 12, 13–20, 72, 298
  affordances *vs.*, xiv–xv, 14, 18, 19
  applied to everyday objects, 132–141
  to bridge Gulf of Execution, 40
  doors and, 15, 16, 132–135
  external, 15
  gesture-controlled devices and lack of, 115–116
  as knowledge in the world, 79
  misleading, 18–19
  perceived affordances and, 145
  poka-yoke technique and, 193
  as reminders, 108–109
  sound as, 155–161
  on touch screen, 21
Signs, 15, 18, 19
Silence, problems with, 157–161
Simplified models, 100–105
Single-spout, single-control faucet, 153–154
Sink drain, signifiers for, 17
Skeuomorphic, 159
Skill-based behavior, 179, 180, 206–207
Sleep, memory and, 95–96
Sleep deprivation, error and, 210, 211
Sliding doors, 16
Slips, 170–171, 172–173
  action, 171, 173, 174, 194
  capture, 174, 208
  classification of, 173–179
  confirmation messages and, 204–205
  description-similarity, 174, 175
  memory-lapse, 171, 173, 176–177, 195
  minimizing, 206–208
  mode error, 174, 177–179
  See also Error; Mistakes
Smart displays/screens, 121, 265–266. *See also* Touch–sensitive displays/screens
Smart phones, 265
Soap dispensers, gesture-controlled, 115–116
Social interaction, 283–284, 298
Social pressure, accidents and, 186–191
Socrates, 286
Sound, as signifier, 155–161
  for electric cars, 157–161
Sound generators, for feedback, 23–24
Spatial cues, natural mapping and, 115
Specifications, design and correct, 234–235
Specify, as stage of execution, 41
Speech, presenting information via, 201–202
Spiral design, 222. *See also* Iteration in design
Stage gate method of product development, 234–235
Stages of action, 40–44
Stairways of public buildings, lockouts and, 144
Standardization
  of faucet design, 153, 154, 155
  individualization *vs.*, 161

Standards
cultural, 130–132
digital time, 252–254
for electric automobile sounds, 159–160, 161
establishing, 248–249
HDTV, 250–252
international, 131, 248–249
necessity of, 250
Startup companies, failure of, 269–270
Stein, Robert, 289
"Sterile Cockpit Configuration," 200
Stigler's law, 270
Stigma problem, 244–247
Story, conceptual models as form of, 57–59
Stove controls, natural mapping and, 113–115, 116–117, 118
Subconscious, 48, 49. *See also* Cognition; Conscious cognition
behavioral level of processing and, 51–52
human thought as, 44–49
skilled behavior as, 206–207
slips and, 173
Subway train doors, lack of signifiers on, 134–135
Susan B. Anthony coin, 79–80, 81–82
Sustainability, model for, 292–293
Swatch International Time, 254
"Swiss cheese model of how errors lead to accidents," 164, 208–210
Switches. *See also* Controls
airplane landing gear, 135
dead man's, 142–143
description-similarity slips and, 175
light, 20–21, 135–140
wireless, 139
System error, 66
System image, 31–32
Task analysis, 137
Tasks
activities *vs.*, 232–234
technology and changes in, 286, 287–288
Taught helplessness, 63
Technical manuals. *See* Manuals
Technological aids, for knowledge in the head, 112–113
Technology
accommodating human behavior, 68–71
adoption of, 268–269, 271, 272, 274
dependence on, 112–113, 285–287
design and, 257
as driver of change, 267–268, 282, 283–285
empowerment of individuals and, 295–297
enhancing human cognition, 112–113, 285–288
handed-up, 297
interaction design and, 5
interplay with psychology, 6–8
meaning of, 281–282
paradox of, 32–34
precision and use of, 104
product innovation and, 258, 264–268
radical innovation and, 281
reminders and, 109
self-blame and trouble using, 63
skeuomorphic designs and, 159
standardization and, 248–254
substituting for memory, 87
Telephone numbers, remembering, 45, 46
Telephones, 68, 70, 156, 264–266. *See also* Cell phones
Telephonoscope, 270–273
Temperature controls, refrigerator, 28–31
Temperature conversions, 101–102
Tenerife disaster, 186–187

Ten-franc coin, 79–80
Testing, 222, 228–229
Text direction/scrolling, culture and choice of, 120–121
Thermostat
  conceptual model of, 57–59, 68–69, 181–182
  control of refrigerator, 28–31
Things That Make Us Smart (Norman), 112, 288
Three-dimensional television, 252
3-D printers, 267, 296, 297
Three Mile Island nuclear power plant accident, 7
Tillers, 21–22
Time
  Australian Aborigines, conception of, 120
  cultural differences in view of, 118–120
  digital, 252–254
  product development and, 236, 238–239
  Swatch International Time, 254
Time-based reminders, 109
Time stress, as cause of error, 168
Touch-sensitive displays/screens, 21, 140, 268–269. *See also* Smart displays/screens
Toyoda, Sakichi, 165
Toyota Motor Company, 165
Toyota Production System, 192, 193
Traffic conventions, 131–132
Training and licensing, 211
Transactive memory, 111–112
Transistor, 281
Transportation accidents, investigation of, 186–187, 188–189
Turn signal switches, 99–100
Typewriters, 280
  development of keyboard, 274–279
Typing
  knowledge in the world and, 77–78
  touch and gesture-sensitive screens and, 264, 266

Ultra-high definition television, 252
Understanding, design and, 3–4
Understanding action, feedback and, 71–72
Underwater (scuba) diving accidents, 187–188
Undo, 199, 203–205
Universal design, 243–247
University of Toronto, 269
Usability, 117–118, 241, 295
Use, determining how to use object, 38–40
User experience, 10, 233
Users
  conceptual model and, 31–32
  designing for, 240–241
  difficulties using products and, 59–65
  purchaser *vs.*, 117–118
  See also Customers

Vacuum tubes, 281
Valance, emotional, 52
Vegetable peelers, design of, 244–245
Vehicular control, mapping and, 21–22
Video conferencing, 273–274
Videophone, 270–274
Video recording, of test groups, 228–229
Visceral level of processing, 50–51, 54–56
Voyager Books, 289

Walkers, design of, 245
Walking, cell phone usage while, 200
Wallace, Wanda, 83
Warning signals, design of, 201–202
Washer-dryer combination controls, 4

Watches
  digital, 27–28, 33–34
  mode-error slips and, 178
  technology and changes in
    design of wristwatch, 32–34
Waterfall method, 234–236
Wegner, Daniel, 112
Wheelchair, control of, 21
Whitehead, Alfred North, 101
Wikipedia, 112, 270, 297
Wireless switches, 139
"Wizard of Oz" prototype
    technique, 227–228
The Wonderful Wizard of Oz (Baum),
    227
Woods, David, 212
Word-gesture typing systems,
    266
Working memory, 92–95, 102
Wristwatch, 32–34
Writing, 104, 106, 107, 267

Xerox Corporation, 227

Zeitgeist, 260
Zhai, Shumin, 266